KDP-family Single Crystals

The Adam Hilger Series on Optics and Optoelectronics

Series Editors: **E R Pike** FRS, **B E A Saleh** and the late **W T Welford** FRS

Other books in the series

Aberrations of Optical Systems
W T Welford

Laser Damage in Optical Materials
R M Wood

Waves in Focal Regions
J J Stamnes

Laser Analytical Spectrochemistry
edited by V S Letokhov

Laser Picosecond Spectroscopy and Photochemistry of Biomolecules
edited by V S Letokhov

Cutting and Polishing Optical and Electronic Materials
G W Fynn and W J A Powell

Prism and Lens Making
F Twyman

The Optical Constants of Bulk Materials and Films
L Ward

Infrared Optical Fibers
T Katsuyama and H Matsumura

Solar Cells and Optics for Photovoltaic Concentration
A Luque

The Fabry–Perot Interferometer
J M Vaughan

Interferometry of Fibrous Materials
N Barakat and A A Hamza

Physics and Chemistry of Crystalline Lithium Niobate
A M Prokhorov and Yu S Kuz'minov

Laser Heating of Metals
A M Prokhorov, V I Konov, I Ursu and I N Mihǎilescu

The Adam Hilger Series on Optics and Optoelectronics

KDP-family Single Crystals

L N Rashkovich

Department of Physics
Moscow State University

Translated from the Russian by Olga Shlakhova

CRC Press
Taylor & Francis Group
Boca Raton London New York

CRC Press is an imprint of the
Taylor & Francis Group, an **informa** business

Taylor & Francis Group
6000 Broken Sound Parkway NW, Suite 300
Boca Raton, FL 33487-2742

First issued in paperback 2019

ISBN-13: 978-0-367-40302-7

British Library Cataloguing in Publication Data

Rashkovich, L. N.
    KDP-family single crystals.
    1. Crystals. Growth
    I. Title
    548.5

Library of Congress Cataloging-in-Publication Data

Rashkovich, L. N. (Leonid Nikolajevich)
    KDP-family single crystals   L.N. Rashkovich ; translated from the
    Russian by Olga Shlakhova.
        p.   cm. — (The Adam Hilger series on optics and
    optoelectronics)

    1. Crystals—Growth.   I. Title.   II. Series.
    QD921.R38   1991
    548'.5—dc20
                                                    90-23493
                                                    CIP

**Visit the Taylor & Francis Web site at
http://www.taylorandfrancis.com**

and the CRC Press Web site at

Contents

Series Editors' Preface

Optics has been a major field of pure and applied physics since the mid 1960s. Lasers have transformed the work of, for example, spectroscopists, metrologists, communication engineers and instrument designers in addition to leading to many detailed developments in the quantum theory of light. Computers have revolutionised the subject of optical design and at the same time new requirements such as laser scanners, very large telescopes and diffractive optical systems have stimulated developments in aberration theory. The increasing use of what were previously not very familiar regions of the spectrum, e.g. the thermal infrared band, has led to the development of new optical materials as well as new optical designs. New detectors have led to better methods of extracting the information from the available signals. These are only some of the reasons for having an *Adam Hilger Series on Optics and Optoelectronics*.

The name Adam Hilger, in fact, is that of one of the most famous precision optical instrument companies in the UK; the company existed as a separate entity until the mid 1940s. As an optical instrument firm Adam Hilger had always published books on optics, perhaps the most notable being Frank Twyman's *Prism and Lens Making*.

Since the purchase of the book publishing company by The Institute of Physics in 1976 their list has been expanded into all areas of physics and related subjects. Books on optics and quantum optics have continued to comprise a significant part of Adam Hilger's output, however, and the present series has some twenty titles in print or to be published shortly. These constitute an essential library for all who work in the optical field.

Preface

The book is intended for physicists, chemists and engineers who grow single crystals from solutions and employ them in physical research and devices.

Although the book is based on the author's research, it contains review material as well and can be used as a reference book. This is true especially as far as Chapter 1 is concerned. Chapter 1 contains nearly all the available data on the physico-chemical analysis of the systems discussed. Chapter 2 represents contemporary concepts on the crystal growth dislocation mechanism which is common for most crystals grown from low- and high-temperature solutions. The chapter on crystal growing is the shortest. The emphasis made is on the possibilities for the considerable speeding up of technological cycles without a deterioration of the crystal properties.

L N Rashkovich
June 1990

Introduction

About fifty years ago Busch discovered the second ferroelectric crystal, besides the already known rochellesalt, KH_2PO_4—otherwise known as KDP (Busch and Scherrer 1935). That was the beginning of large-scale investigations into the properties of KDP-family single crystals and their commercial applications. These fifty years of investigation are reviewed in two volumes of the journal *Ferroelectrics* (volumes 71 and 72). Let us recall the milestones in the history of these crystals. In 1938 Busch discovered the ferroelectric (and antiferroelectric) properties of KH_2AsO_4, $NH_4H_2PO_4$ and $NH_4H_2AsO_4$. In 1939 Ubbelohde discovered the ferroelectric phase transition in KD_2PO_4 (DKDP). In 1944 Zwicker and Sherrer found that KDP and DKDP possessed a significant electro-optic effect. The crystalline structure of KH_2PO_4 was described by West (1930) and then in 1953 Bacon and Pease determined the position of hydrogen atoms in this structure. The first theory of the ferroelectric phase transition in KDP was suggested by Slater in 1941.

In the 1930s to 1940s, crystals of $NH_4H_2PO_4$ (ADP), which is a crystallochemical analogue of KDP, found a fairly wide practical application. They were used as sound piezoelectric transducers in microphones, gramophones and other sound reproduction devices.

A new impetus was given to the production of the KDP-family crystals (especially KDP and DKDP) soon after the advent of lasers, when for the first time, the second harmonic of the fundamental radiation was produced at the KDP crystal using the effect of phase synchronism (Giordmaine 1962). At the present time, KDP-family crystals are the primary materials used in devices controlling laser radiation, its modulation, frequency conversion and scanning. They are manufactured in larger quantities than the sum total of all other crystals used in quantum electronics and all crystals grown from solutions in all their possible applications.

Hundreds of papers are devoted to investigation of the physical properties of the KDP-family crystals (with a general formula $(Me_1,Me_2)(H,D)_2(P,As)O_4$) and the number of such works does not diminish. The

1

majority of the known data is represented in Landolt-Börnstein (1984) and in a number of reviews. For instance, Nelmes (1987) gave a summary of the structural studies, Eimerl (1987a) generalized the studies of electro-optical, linear and non-linear optical properties. The review by Courtens (1987) is devoted to mixed crystals of the spin-glass type. The theory and peculiarities of phase transitions in the KDP-family crystals are treated in more than a dozen monographs.

This book is devoted to the physical and physico-chemical problems related to growing the KDP-family crystals. Although these questions are dealt with in a great number of publications, the problem is still very important and requires further investigation because of the growing demands in optical perfection and sizes of crystals. The developing power optics need single-crystal elements with an aperture of ~ 100 cm—a fantastic size for artificial single crystals. Nevertheless, it is possible to grow such crystals. In power optics devices, powerful laser beams are being transmitted and even a slight absorption or scattering of light in a crystal may destroy the material or lead to an unacceptable distortion of the radiation characteristics. To preserve the high quality of large crystals is a more difficult task than that of increasing the sizes of crystals. But certain progress has been attained in solving this problem too. I hope the book will be useful for both industrial and research workers who study the problems of the growth and application of single crystals.

1 Phase Diagrams of the $Me_2O-P_2O_5(As_2O_5)-(H,D)_2O$ Systems in the Crystallization Region of Mono-substituted Salts

The study of phase diagrams of physico-chemical systems in which crystallization is taking place and the investigaton of phases coexisting in such systems provide the basis for improving the growth techniques and perfecting the optical properties of crystals. As far as the $Me_2O-P_2O_5(As_2O_5)-(H,D)_2O$ systems are concerned little research appears to have been done until now and there are still a considerable number of questions to be answered, particularly concerning heavy water systems.

There is in addition another important aspect of phase diagrams which deserves special attention. Substituting deuterium for hydrogen and altering an alkaline cation results in regular changes in the properties of solid and liquid phases. These problems are of some importance for both material science and solid state physics.

In the 1930s academician N S Kurnakov and his co-workers performed classical investigations of the $(NH_4)_2O-P_2O_5-H_2O$ system and particularly the $K_2O-P_2O_5-H_2O$ system. Thirty years later Barkova and Lepeshkov (1966, 1968) studied the $K_2O-P_2O_5-D_2O$ system, and Rashkovich et al (1967) examined the crystallization conditions and properties of $ND_4D_2PO_4$ solutions. Later the author studied similar systems with rubidium and caesium. The available material has been partially compiled by Eysseltova and Dirkse (1988).

In this chapter the systems in question are dealt with in turn. First, for each system the crystalline phases of mono-substituted salts followed by the

3

dependence of the degree of deuteration of the solution on the temperature region of spontaneous crystallization are discussed. Then the data on solubility polytherms and isotherms and those on solution properties (density, conductivity, etc) are given. The last sections describe the characteristic properties of liquidus surfaces of the corresponding diagrams, general features of their topology, and the influence of kinds of cation and solvent on the topology of phase diagrams. Finally, isotopic exchange equilibrium in the liquid phase as well as between a crystal and solution is considered.

Note that in the first sections the deuteration degree, x, is taken to be the solution's deuteration degree (the ratio of heavy-water molar content to the total amount of water), unless stated otherwise.

1.1   The $[N(H,D)_4]_2O-P_2O_5-(H,D)_2O$ system

The first system to be considered is the $(NH_4)_2O-P_2O_5-H_2O$ system. This system was studied at 0, 25 and 50°C (Muromtsev and Nazarova 1938) and at 25°C (Flatt *et al* 1951). Earlier studies at 25°C are discussed by Volfkovich *et al* (1932). At 75°C the system was examined by Brosheer and Anderson (1946), and at 100°C by Wenyu (1985). Many works only described the temperature dependence of the $NH_4H_2PO_4$ solubility in water and in solutions of differing acidity. Apparently, data on the crystallization of $ND_4D_2PO_4$ were reported solely by Rashkovich *et al* (1967) and Vasilevskaya *et al* 1967.

1.1.1   Solid phase
In the whole existence region $N(H,D)_4(H,D)_2PO_4$ crystallizes only as tetragonal crystals, this habit being typical of a KH_2PO_4 crystal group. Changing the pH of the crystallization medium does not lead to the formation of new faces but affects the rate of face growth of prisms and dipyramids. Byteva (1962, 1965, 1966) showed that as pH increases from 3.5 to 6.3 this rate increases monotonically. Similar research was later carried out in more detail by Davey and Mullin (1976b). Impurities of trivalent metals combined with supersaturation, pH, temperature and stirring conditions may result in the so called tapering of crystals, i.e. non-parallel growth of prismatic faces, with the angle between them ranging from 0 to ≈ 30°C (Davey and Mullin 1974a, 1974b, 1976a, Loiacono *et al* 1982, Takubo *et al* 1984, Dam *et al* 1986). This phenomenon is also typical of other crystals of the KH_2PO_4 group. The nature of crystal morphology changes are not well understood. Thus, for instance, Aguilo and Woensdregt (1987) analysed variants of the possible equilibrium form of crystals and came to the conclusion that changing the ammonium content in solution cannot result in morphology changes, so the reason for this effect has to be sought elsewhere.

Napijala *et al* (1978) found the Curie temperature and other properties of $NH_4H_2PO_4$ crystals to depend on the pH of the solution they were grown from. They supposed that in an acid medium several NH_4^+ groups could be replaced by H_3O^+. Replacing hydrogen by deuterium is easier in an NH_4 group. This problem has not been studied in detail, however, but it is known that with the total deuteration degree $x > 0.9$ the replacement in an NH_4 group is complete, whereas in the anion the deuteration degree is less than the total one (Fukami *et al* 1986).

1.1.2 Solubility polytherms

The solubility of NH_4H_2PO in water was studied by Buchanan and Winner (1920), Volfkovich *et al* (1932), Bergman and Bochkarev (1938), Polosin (1946), Polosin and Treshchov (1953) and others. The results obtained are in good agreement and can be approximated quite well by linear relations if the values of the solubility c are taken in mass per cent. For example, Dauncey and Still (1952) gave the formula

$$c = 17.2 \mid 0.474\,t.$$

In 1967 Mullin and Amatavivadhana determined the solubility in the temperature range 20–40°C. For these results the author and co-workers obtained the following relation

$$c = (16.73 \pm 0.35) + (0.484 \pm 0.012)\,t \pm 0.18. \tag{1.1}$$

The temperature dependence can be expressed by a more accurate though less convenient relation

$$\ln m = A + B/T + D\ln T$$

where m is the mole fraction of the salt and T is the absolute temperature. The following values are given (Broul *et al* 1979) for the coefficients

$$A = -11.9204 \qquad B = -1139.73\ \mathrm{K} \qquad D = 2.2684$$

or (Vogel *et al* 1983) for $t = 30$–60°C

$$A = 36.422 \qquad B = -3370.284\ \mathrm{K} \qquad D = -4.901.$$

All these relations differ at most by 0.2 wt %.

Heating $NH_4H_2PO_4$ in air results in the decomposition of the salt. This process begins at $t < 130$°C (after one hour at 130°C, less than 5 wt % of the salt has been transformed into pyrophosphate $(NH_4)_2H_2P_2O_7$; at 170°C, the transformation into pyrophosphate is complete (Rilo and Kulikov 1981)). The maximum rate of mass loss in a non-deuterated crystal is observed at ≈ 225°C, and with $x = 95\%$ at 235°C. Upon decomposition of ADP, the NH_3 group is released at a slightly lower temperature than H_2O.

If the heating is being performed in a closed container, these transformations are reversible; thus Pastor (1985), for example, melted $NH_4H_2PO_4$ at 215°C (with a pressure < 200 atm) in a closed quartz ampoule and upon cooling no transformations of the phase composition were observed. This fact led Pastor to suggest a technique of growing crystals from the melt.

The solubility of $ND_4D_2PO_4$ in D_2O (with the deuteration degree of solution being $\approx 98\%$) was measured by the author using the concentration flow technique (Mullin 1972, p 42). The following data were obtained

Salt concentration (wt %)	35	40	44.4	50	55
Saturation temperature (°C)	20	30.4	41.1	52.9	64.8.

These results can also be approximated by a linear dependence

$$c = (26.2 \pm 0.3) + (0.446 \pm 0.006)t \pm 0.2. \qquad (1.2)$$

1.1.3 Solubility isotherms
The solubility isotherms of $NH_4H_2PO_4$ in the $(NH_4)_2O-P_2O_5-H_2O$ system at 25°C and 50°C are given in figure 1.1(a), according to the data of Muromtsev and Nazarova (1938) and of Flatt *et al* (1951). At 25°C both branches are, in fact, linear; the solubility minimum is singular and corresponds to a stoichiometric solution. The linear dependence constant of the isotherm branches are given in table 1.16. The acid branches of the isotherms intersect the straight line corresponding to the molar ratio of the solution $(NH_4)_2O/P_2O_5 = 1/2$, consequently, the ultraacid salt $NH_4H_5(PO_4)_2$ dissolves in water incongruently. At 75°C and 100°C the isotherms have the same shape.

The correlation between the concentration of solutions saturated at 30°C and their pH was studied by Byteva (1968).

The ice crystallization field in this system was examined by Kurnakov *et al* (1938). Figure 1.1(b) shows that in this region there is a distinct kink corresponding to solutions with molar ratio $(NH_4)_2O/P_2O_5 = 1$.

1.1.4 Properties of solutions

(a) *Density*
The concentrational dependence of the density of $NH_4H_2PO_4$ solutions with stoichiometric composition at 23°C was investigated by Chomjakow *et al* (1933). The analysis of their data yields the following dependence

$$d = 0.9965 + 0.0058c \pm 0.0005 \text{ (g cm}^{-3}). \qquad (1.3)$$

Mullin and Amatavivadhana (1967) determined solution densities at three temperatures. Their data can be generalized by the following relations:

20°C: $d = (0.9982 \pm 0.0008) + (0.00580 \pm 0.00005)c \pm 0.0009$

30°C: $d = (0.9962 \pm 0.0011) + (0.00579 \pm 0.00006)c \pm 0.0011$ (1.3a)

40°C: $d = (0.9921 \pm 0.0010) + (0.00586 \pm 0.00006)c \pm 0.0010.$

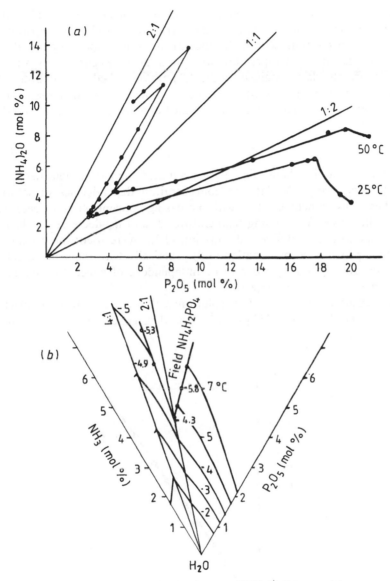

Figure 1.1 (a) The solubility isotherms of $NH_4H_2PO_4$: ●, Muromtsev and Nazarova (1938); ○, Flatt *et al* (1951); and (b) the crystallization isotherms of ice (Kurnakov *et al* 1938).

From these data and equation (1.1), we determine the saturated solution density to be

$$d_{sat} = 1.093 + 0.00282t \pm 0.003. \tag{1.4}$$

We measured the density of solutions with the salt concentration varying from 9 to 50 wt %. The experiments were carried out at different temperatures (25–75°C). In solutions with high concentrations, a temperature about 5°C above the saturation temperature was maintained. Regardless of varying temperature, the data obtained (figure 1.2) are described by equation (1.3).

If concentration is expressed in volume fractions, $c^* = cd$, then the solution density can be expressed in terms of the partial density of water d_w and that of the salt d_s

$$d = d_w + (1 - d_w/d_s)c^*. \tag{1.5}$$

Fedotova and Tsekhanskaja (1984) calculated the partial densities using, unfortunately, very old data on concentration dependencies of solution densities. They concluded that, with increasing concentration, d_s decreases and d_w increases. With rising temperature, d_s and d_w decrease (table 1.1). The result would be different if equation (1.3a) were used; these relations can be approximated by equation (1.5) practically with the same standard deviation and d_s and d_w prove to be independent of concentration. These results are also given in table 1.1.

The density of an $ND_4D_2PO_4$ solution with a concentration of 40 wt % measured at 35°C is $1.334\,\mathrm{g\,cm^{-3}}$. The density of D_2O at the same

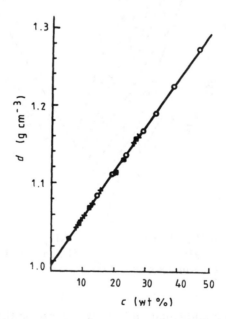

Figure 1.2 The solution density of $NH_4H_2PO_4$: +, Chomjakow *et al* (1933); □, Mullin and Amatavivadhana (1967) at 30°C; ○, Rashkovich *et al* (1967).

Table 1.1 Partial density of water and $NH_4H_2PO_4$ in solution in g cm^{-3} (crystal density 1.792 g cm^{-3}).

(a) From Fedotova and Tsekhanskaja (1984)

		25°C		60°C	
m	c (wt %)	d_s	d_w	d_s	d_w
0	0	0.998	2.15	0.981	2.12
0.02	11.5	1.000	2.04	0.982	2.01
0.04	21.0	1.005	1.94	0.987	1.92
0.06	29.0	1.013	1.85	0.995	1.83
0.08	35.0	—	—	1.007	1.75

(b) From Mullin and Amatavivadhana (1967) using equation (1.3a)

t (°C)	d_w	d_s	$B = 1 - d_w/d_s$
20	1.005 ± 0.007	1.94 ± 0.02	0.484 ± 0.001
30	1.003 ± 0.006	1.95 ± 0.02	0.486 ± 0.001
40	0.999 ± 0.005	1.97 ± 0.01	0.493 ± 0.001

temperature is 1.100 g cm^{-3}. Assuming the concentration dependence of the density to be linear we obtain

$$d = 1.100 + 0.00585c. \tag{1.6}$$

(b) *Refraction indices*
The refraction indices of solutions were measured at 40°C under a sodium lamp. The experimental data are described by linear dependences (Rashkovich *et al* 1967)

$$n_{NH_4H_2PO_4} = (1.3311 \pm 0.0002) + (0.001371 \pm 0.000009)c \pm 0.0003$$

$$n_{ND_4D_2PO_4} = (1.3276 \pm 0.0006) + (0.00140 \pm 0.00003)c \pm 0.0008.$$

$$\tag{1.7}$$

When the temperature of solutions with constant concentration was changed by 10°C, the refraction index changed by 0.0015 for H_2O solutions and by 0.0010 for D_2O solutions (Rashkovich *et al* 1967). According to Takubo and Makita (1989), at 40°C $n_{NH_4H_2PO_4} = 1.3305 + 0.0013761c$. These authors obtained similar relations at 20°C and 60°C and give the temperature dependence for saturated solutions (at 0–70°C) as

$$n_{NH_4H_2PO_4} = 1.3608 + 0.00048731t.$$

(c) *Conductivity*

Watkins and Jones (1915) measured the conductivity and dissociation degree of $NH_4H_2PO_4$ in aqueous solutions with concentrations less than 2.8 wt % at 0–35°C. In such solutions the dissociation degree did not actually depend on temperature and amounted to 66% at maximum concentration. The specific conductivity of solutions with constant concentration depended linearly on temperature.

Rashkovich *et al* (1967) made measurements at a frequency of 1 kHz with an accuracy of ± 0.2%. The accuracy of the temperature measurements was ± 0.05°C, that of the concentration ± 0.1 wt%. Some of these experimental data are shown in figure 1.3. One can see that the conductivity (χ) of solutions with constant concentration increases linearly with rising temperature. Constants of the corresponding equations are given in table 1.2.

Empirical dependences of the conductivity are found to be in good agreement with the data for the constants *a* and *b* given in table 1.2. These dependences are:

for H_2O solutions,

$$a/c = 0.3961 - 0.04415\sqrt{c} \qquad b/c = 0.01432 - 0.001487\sqrt{c} \qquad (1.8)$$

for D_2O solutions,

$$a/c = 0.2485 - 0.0251\sqrt{c} \qquad b/c = 0.009495 - 0.0008011\sqrt{c}. \qquad (1.9)$$

Extrapolation of the author's data over the range of low concentrations agrees with the results obtained by Watkins and Jones (1915). These authors

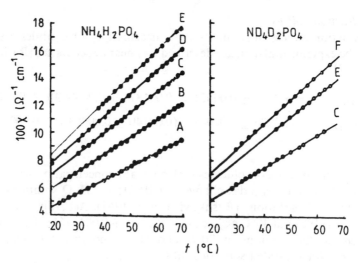

Figure 1.3 Conductivity of $N(H,D)_4(H,D)_2PO_4$ solutions of different concentration, *c* (wt %): A, 10; B, 15; C, 20; D, 25; E, 30; F, 39.

Table 1.2 Constants of the equation $100\chi = a + bt$. The temperature range is 25–70°C, χ is in Ω^{-1} cm^{-1}, t is in °C.

c (wt %)	a	b
(a) H$_2$O solutions		
8.6	2.31	0.0853
10	2.56	0.0991
15	3.34	0.1260
20	3.94	0.1500
25	4.43	0.1685
30	4.67	0.1890
35	4.75	0.1950
40	4.62	0.1990
(b) D$_2$O solutions		
20	2.72	0.1183
30	3.36	0.1533
39	3.55	0.1750

calculated the value of the molar conductivity as $\lambda = 66.66\,\Omega^{-1}$ cm^2 at 25°C for solutions of 0.25 molar concentration ($c = 2.83$ wt %). Calculation by empirical formulae (equations (1.8) and (1.9)) yields $\lambda = 66.70\,\Omega^{-1}$ cm^2.

Increasing the solution concentration at constant temperature leads first to an increase in χ and then, as a result of decreasing ion mobility and dissociation degree, χ diminishes. The concentration corresponding to maximum conductivity can easily be found by differentiating the equation for χ with respect to concentration. For H$_2$O solutions this concentration c_{max} is 38.1 wt % at 25°C and 39.3 wt % at 60°C. For D$_2$O solutions at the same temperature the maximum values of χ are 51.6 and 55.8 wt % of ND$_4$D$_2$PO$_4$, respectively. The fact that these values are higher than in the case of NH$_4$H$_2$PO$_4$ indicates a stronger interaction of dissolved particles in H$_2$O. The weak temperature effect on the concentration value corresponding to maximum conductivty indicates (in agreement with the results obtained by Watkins and Jones (1915)) the weak temperature effect on the dissociation degree.

Due to a lower mobility of D$^+$ ions as composed to H$^+$ ions, the specific conductivity of ND$_4$D$_2$PO$_4$ is less than that of NH$_4$H$_2$PO$_4$ solutions (c and t being equal). Therefore, regardless of the higher solubility of ND$_4$D$_2$PO$_4$, the conductivity of saturated solutions of this salt is less than in the case of NH$_4$H$_2$PO$_4$ (figure 1.4). Molar conductivity of 0.25 mol ND$_4$D$_2$PO$_4$ at 25°C is 48.04 Ω^{-1} cm^2 as compared to 69.70 Ω^{-1} cm^2 for

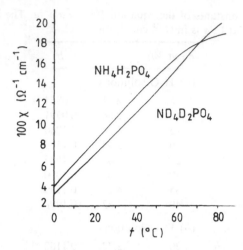

Figure 1.4   Conductivity of saturated solutions of $N(H,D)_4(H,D)_2PO_4$.

$NH_4H_2PO_4$ solutions under the same conditions. The ratio of these values (0.689) is close to that of D^+ and H^+ mobility (0.676) at the same temperature.

These results could prove to be useful for continuous control of the solution concentration (at constant temperature) when single crystals are being grown. Figure 1.5 represents the exemplifying graphs calculated for this purpose.

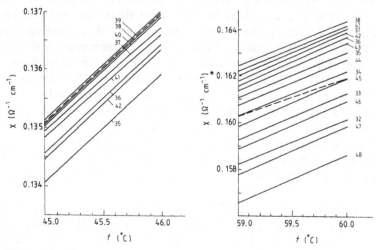

Figure 1.5   Temperature dependence of the conductivity of $NH_4H_2PO_4$ solutions. Concentrations (wt %) are denoted beside the curves. The broken lines refer to the saturated solution.

One can see that the greater the difference between the solution concentration and the concentration corresponding to maximum conductivity, the greater is the variation of χ with the equivalent variation of concentration. Differentiating the expression for χ with respect to c and considering equation (1.8) we obtain

$$d\chi/dc = 1.5 \times 10^{-2}(0.04415 + 0.001487t)(\sqrt{c_{\text{max}}} - \sqrt{c}).$$

At 60°C the solution concentration of $NH_4H_2PO_4$ with our measurement accuracy could be determined with an error of ± 0.1 wt %. At lower temperatures, $d\chi/dc$ diminishes due to decreasing values of both factors in the above expression (see also figure 1.5), and so the accuracy of the concentration determination decreases. In the case of $ND_4D_2PO_4$ the situation is even worse, since $d\chi/dc$ is a smaller value.

(d) *Diffusion and viscosity*
Diffusion and viscosity were studied by Mullin and Cook (1963). In the temperature range 15–50°C these quantities are related by the relation

$$D\eta \times 10^8 = 4.7 + 0.16t$$

where D is in cm^2 s^{-1}, η is in poise and t is in °C. The viscosity is characterized by the following values (in centipoise)

c (wt %)	20°C	30°C	40°C
0	1.005	0.801	0.656
16.7	1.590	1.271	1.042
23.2	2.035	1.610	1.311
27.8	2.476	1.942	1.548

(e) *Surface energy*
The surface energy, α, of the interface between a crystal and solution, according to data on crystallization kinetics (see Chapter 2) and those obtained by Rashkovich *et al* (1985) for growth steps at a prismatic face, is 16–20 erg cm^{-2} (at ~ 35°C). Lower values (4–8 erg cm^{-2}) are obtained in experiments on spontaneous nucleation which is assumed to be homogeneous (Nagalingam *et al* 1980, 1981). It appears that underestimated values of α are due to the nucleation being in fact heterogeneous.

Having analysed the data reported by other authors for more than fifty aqueous systems and using the empirical relation, Söhnel (1982) obtained $\alpha = 20$ erg cm^{-2} for $NH_4H_2PO_4$.

1.2 The $K_2O-P_2O_5-(H,D)_2O$ system

The system in question is in many respects similar to that discussed above. The only essential difference is that, instead of one, two crystalline modifications of mono-substituted orthophosphate are formed from deuterated solutions. Most investigations are devoted to the crystallization conditions of KH_2PO_4, while the crystallization of KD_2PO_4 has been studied much less extensively.

1.2.1 Solid phases

It is only the tetragonal phase that is crystallized from H_2O solutions of any acidity at temperatures ranging from that of the eutectic up to at least 100°C. The situation is the same in deuterated solutions with a deuteration degree $x < 60\%$. As in the case of $NH_4H_2PO_4$, increasing the pH of the solutions slightly affects the habit of the crystals, i.e. the growth rate of the prismatic faces increases and dipyramids become smaller (see, for example, Gavrilova and Kuznetsova 1966 and Franke *et al* 1975). Tapering occurs when impurities of heavy metals are present.

The structures of tetragonal KH_2PO_4 and KD_2PO_4 crystals differ by a slight relative rotation of neighbouring PO_4 tetrahedrons leading to an increase of the hydrogen bond length, with hydrogen being substituted by deuterium. The other difference between the two structures is a different (although slight) deviation of a proton (a deuteron) from the line connecting the nearest oxygen atoms of neighbouring tetrahedrons (Nelmes *et al* 1972, Nakano *et al* 1973, Kennedy *et al* 1976). The density of KH_2PO_4 crystals is 2.335 g cm^{-3}, that of KD_2PO_4 crystals is 2.359 g cm^{-3}.

In 1939, Ubbelohde found that the phase with monoclinic symmetry crystallizes from solutions with a deuteration degree of about 100%. Small crystals of this phase quickly became opaque in the air and in a few days transformed into the tetragonal phase (Ubbelohde and Woodward 1939, 1942).

When tetragonal KD_2PO_4 crystals are grown with the maximum degree of deuteration, spontaneous crystallization of the monoclinic phase hinders this process, since in this case tetragonal crystals not only stop growing but can dissolve completely.

The monoclinic phase structure was first studied in 1972 (Nelmes 1972) and was specified in 1975 (Thornley and Nelmes 1975). The crystals have the spatial symmetry group C_2^2–$P2_1$, the only symmetry element of which being the second-order polar axis. The habit of the crystals has not been described in the literature. Figure 1.6 presents a photograph of a crystal aggregation of both phases. Our observations of a large number of monoclinic crystals grown under different conditions showed the crystals to have a lengthened prismatic habit, the symmetry axis being perpendicular to the long crystal face. The most developed crystal faces are those of front pinacoid {100}, right- and left-side monohedrons (010) and (0$\bar{1}$0) and longitudinal prismatic faces {011}. In addition, the following simple forms are also presented: vertical prismatic faces {1$\bar{1}$0}, {130}, {120}, a longitudinal prismatic face {012} and an axial dihedron {1$\bar{1}$1}. Most crystals observed referred to left-handed enantiomorphic forms.

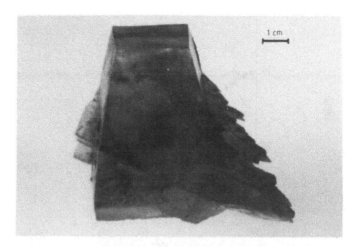

Figure 1.6   The aggregation of KD_2PO_4 monoclinic crystals intergrown through the tetragonal crystal. Pyramids of the tapered tetragonal crystal and the left-hand part of monoclinic ones are cut off. The difference in solubility of both phases is so small that the habit of the more soluble tetragonal crystal is preserved.

The freshly polished surface of crystals is quickly covered by a film of the tetragonal phase. Centimetre monoclinic crystals with a deuteration degree of 80% transform completely into the tetragonal modification during one year. Samples of the same dimensions with a deuteration degree of 95% are covered with a thin film of the tetragonal phase over the same time period. Monoclinic-phase powder with a deuteration degree of 95% completely transforms into the tetragonal modification in air in about one hour. Taking a Debyegram immediately after grinding in air, it is possible to detect only three to four strong monoclinic phase lines and five to eight weak ones.

The first study of optical, electro-optical and non-linear optical properties of monoclinic crystals was reported by Alexandrovskii *et al* (1978) (see also Jamroz and Górsi (1979)). Dielectric properties were studied by Blinc *et al* (1975) and Yakushkin and Anisimova (1988).

Data on monoclinic-phase spontaneous crystallization were reported by Belouet *et al* (1975). When growing tetragonal $K(H,D)_2PO_4$ crystals the authors observed a prolonged simultaneous coexistence of both phases under conditions represented by the curve in figure 1.7: the tetragonal phase is stable below the curve, the monoclinic one is stable above the curve. The solutions were of pH = 4.45.

Strel'nikova and Rashkovich (1977) investigated spontaneous crystallization with slow evaporation of the solvent in the temperature range 5–70°C, pH = 1.6–4.2 and with different deuteration degrees of the solution. The results are given in figure 1.8. The curves drawn through the experimental

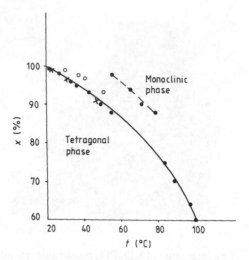

Figure 1.7 Existence regions of monoclinic and tetragonal phases in $K(H,D)_2PO_4$ solutions: ○, Belouet *et al* (1975); ●, Jiang *et al* (1981); ×, Batyreva *et al* (1981). The broken curve refers to the solid-phase transition of tetragonal crystals into monoclinic ones (in solution).

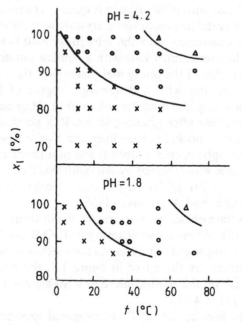

Figure 1.8 The spontaneous crystallization regions of solid phases in $K(H,D)_2PO_4$ solutions: ×, the tetragonal phase; △, the monoclinic phase; ○, a mixture of both phases.

points determine three crystallization regions: those of tetragonal phase, monoclinic phase and the mixture of both. The spontaneous crystallization region of one monoclinic phase is determined by a small number of points, since above 70°C evaporation of the solvent proceeded quickly; crystals were small and it was rather difficult to identify them.

One can see in figure 1.8 that with $x < 0.8$ the monoclinic phase does not crystallize spontaneously. In solutions with a greater degree of deuteration the position of the spontaneous crystallization upper boundary of one tetragonal phase in solution with pH = 1.8 is appreciably shifted into the higher temperature region as compared with a solution with pH = 4.2. Thus, with $x = 0.99$ in a solution with pH = 4.2 the monoclinic phase does not emerge spontaneously at temperatures of about 0°C. However, decreasing the pH to 1.8 makes it possible to raise the temperature by 10–15°C.

The characteristic feature of the results obtained is a broad temperature region of simultaneous crystallization for both phases. It proves that the solubility of solid phases in the temperature region in question is rather close and the solution in this region is supersaturated with respect to both phases.

To determine the temperature of the intersection of the solid phases' solubility curves the temperature dependences of solubility were measured for either phase. The concentration flow technique was used. Tetragonal and monoclinic crystals were in turn immersed into solution and then removed whilst keeping the measuring cell hermetically sealed.

Solutions were prepared in such a way that the free acid content (calculated per solvent) and the deuteration degree were independent of the salt concentration. To achieve these conditions the salt was dissolved completely by heating the solution up to $\simeq 60°C$ with subsequent cooling to room temperature. In these experiments the container with the solid and liquid phases was kept for a few hours at a temperature close to the required one; then the liquid phase was decanted, superheated and poured into a measuring cell. The solution concentration was determined after measuring the saturation temperature. Note that solutions, prepared in this manner and having pH = 4.2 at room temperature, contained 0.4 wt % of redundant K_2O (in relation to the solvent); while solutions with pH = 1.6 contained 21.35 ± 0.10 wt % of $(D,H)_3PO_4$ (also in relation to the pure solvent).

The deuteration degree of the crystals used was 0.87 (for the tetragonal crystal) and 0.96 (for the monoclinic one). Some difference in the deuteration degree of crystals and solution would not alter significantly the results obtained; since the surface crystal layer and solution have reached isotopic equilibrium due to isotopic exchange during repeated dissolution and crystallization. It seems to be essential that the monoclinic phase eventually transforms into a tetragonal one. This process is particularly rapid in an acid medium, moreover, it accelerates with lowering deuteration degree. For this reason crystals were carefully examined after each measurement. The transformation into the tetragonal phase was detected by the loss of transparency. When this occurred the experiment was repeated with new samples.

The accuracy of the deuteration degree determination was ± 0.005, that of concentration $\pm 0.3-0.5$ wt %, and that of the saturation temperature $\pm 0.3\,°C$.

The data obtained were analysed by the least-squares technique under the assumption of a linear dependence of solubility with temperature, the results being represented in table 1.3. The table shows that the mean-square error of the linear approximation was of the same order as the accuracy of the concentration determination.

The data reported by Strel'nikova and Rashkovich (1977) can be compared with the results obtained by Barkova and Lepeshkov (1968) for the tetragonal phase with $x = 98\%$. These authors determined the solubility of the tetragonal phase to be equal to 26.8 wt % in a solvent containing 0.23 wt % of K_2O (our result being 25.7 wt % in a solvent containing 0.4 wt % of K_2O) and to equal 31.4 wt % in a solvent containing 14.9 wt % of P_2O_5 (our result being 30.8 wt % in 15.0 wt % of P_2O_5).

The analysis of the data represented in table 1.3 reveals only a slight difference in the solubility of tetragonal and monoclinic crystals. For example, this difference does not exceed 0.5 wt % at $\pm 20\,°C$ from the intersection point of the solubility isotherms for all solutions. This is due to the fact that the research techniques available at present do not allow the determination of the temperature of the curve intersection with sufficient accuracy (refer to the right-hand column in table 1.3).

Jiang et al (1981) carried out a detailed study of the relative solubilities being discussed. Solutions with pH $= 3.1-3.3$ were used. The concentration flow technique was also used to determine the saturation temperature. The accuracy of the determination of the solubility curve intersection for both

Table 1.3 Constants of solubility equations ($c = a + bt$) for the tetragonal and monoclinic phases of $K(D_xH_{1-x})_2PO_4$.

| x | pH | Tetragonal phase | | | Monoclinic phase | | | Inter-section tempera-ture (°C) |
		a (wt %)	b (wt %/°C)	σ_0	a (wt %)	b (wt %/°C)	σ_0	
0.98	4.2	16.7	0.360	0.26	17.5	0.330	0.20	27 ± 15
	1.6	23.7	0.285	0.20	24.2	0.266	0.17	26 ± 19
0.86	4.2	15.6	0.366	0.36	16.6	0.351	0.28	67 ± 43
	1.6	22.6	0.300	0.31	23.5	0.281	0.06	47 ± 19
0.77	4.2	14.9	0.346	0.46	16.9	0.310	0.51	56 ± 27
0.996	3.2	17.98	0.333	—	19.07	0.283	—	21.6
0.889	3.2	—	—	—	18.3	0.33	0.80	—

phases is estimated by the authors to be $\pm(0.5-2)°C$. Their results are represented in figure 1.7. The temperature dependences of the solubility are not given by the authors. Recalculation of the results provided by the authors in their graphs and tables are given in table 1.3 (the last two lines).

Table 1.3 demonstrates the qualitative agreement between the results; however, there is a certain divergence with the data given in figure 1.7, the main difference being that the curve of Jiang *et al* (1981) is below that of Belouet *et al* (1975) regardless of the lower value of pH. At the same time there is no doubt as far as the prevention of spontaneous production of monoclinic crystals in an acid medium is concerned. This fact accounts for using a solution with pH = 2 in the commercial growth of KD_2PO_4 tetragonal crystals when crystallization is performed under decreasing temperature in the range 25–12°C (Loiacono 1975, 1987).

The results of Batyreva *et al* (1981) occupy the intermediate position between those of Belouet *et al* (1975) and Jiang *et al* (1981). The former are also given in figure 1.7. In those experiments the solution used was of pH $\simeq$ 4.

Tetragonal crystals with $x < 98\%$ can be grown in the region above the solubility curve of the monoclinic phase until (if) nucleation centres of the latter emerge spontaneously. However, just as monoclinic crystals in solution transform into tetragonal ones, so tetragonal crystals transform into monoclinic ones above a certain temperature. This solid-phase transition temperature was determined by Jiang *et al* (1981) and is shown by the broken curve in figure 1.7. Thus, growing tetragonal crystals becomes possible only below the curve mentioned (with pH = 3.1–3.3).

1.2.2 Solubility polytherms

The solubility of KH_2PO_4 in water has been determined many times. All results are in good agreement with each other and with the empirical formula derived by Kazantsev (1938) for temperatures in the region 10–90°C:

$$c = 12.79 + 0.250t + 0.00182t^2 - 0.00000616t^3.$$

For the temperature range 20–60°C, from the same data we obtain

$$c = (10.9 \pm 0.2) + (0.364 \pm 0.005)t \pm 0.2.$$

Recalculation of the data by Mullin and Amatavivadhana (1967) for the 20–40°C region gives

$$c = (11.6 \pm 0.3) + (0.335 \pm 0.011)t \pm 0.2.$$

For 16 measurements in the temperature range 29–50°C Rosmalen (1977) obtained a more accurate relation

$$c = 10.68 + 0.3616t \pm 0.04. \tag{1.10}$$

In the mole fraction scale the temperature dependence of solubility is expressed, according to Broul *et al* (1979), in the form

$$\ln m = -14.756 - 1043.492/T + 2.599 \ln T$$

and, according to Vogel *et al* (1983), in the form

$$\ln m = -4.810 - 1510.549/T + 1.128 \ln T$$

All these relations agree with each other within an accuracy of not less than ± 0.2 wt %.

The $KH_2PO_4-H_2O$ system was investigated by Marshall (1982) at 275–400°C. It was found that two liquids are produced above 385.6°C (with $\simeq 12$ wt % of KH_2PO_4 at the lower critical point).

The solubility of KD_2PO_4 tetragonal crystals in D_2O ($x \simeq 0.98\%$) was studied by Barkova and Lepeshkov (1966) in the range 0–100°C. From their data we obtain the following relation for the 20–80°C temperature region

$$c = (17.89 \pm 0.07) + (0.3531 \pm 0.0015)t \pm 0.06. \qquad (1.11)$$

Similar dependences obtained in our experiments and in those of Jiang *et al* (1981) are represented in table 1.3. In solutions, irrespective of the deuteration degree, the solubility (in wt %) depends linearly on temperature. Havránková and Březina (1974) showed that at 30°C, 45°C and 60°C in a solution of pH $\simeq 3.5$ (the solution composition being different from the stoichiometric composition), the solubility (in g KH_2PO_4 per 100 g H_2O) depends linearly on the deuteration degree.

When heated in a sealed container, KH_2PO_4 melts congruently at 272°C (Zhigarnovskii *et al* 1984, Pastor and Pastor 1987). Heating in air leads to the gradual removal of water and the production of highly soluble polyphosphates. This fact allows us to produce a highly deuterated solution by dissolving dehydrated KH_2PO_4 in D_2O. The process for producing polyphosphates is

$$KH_2PO_4 \Rightarrow K_2H_2P_2O_7 \Rightarrow K_3H_2P_3O_{10} \Rightarrow \ldots \Rightarrow (KPO_3)_n.$$

Shchegrov *et al* (1982) determined the composition of the thermal dehydration products upon heating for 30 min (in wt %):

	250°C	350°C
KH_2PO_4	26.2	16.1
$K_2H_2P_2O_7$	15.3	19.3
$K_3H_2P_3O_{10}$	11.2	18.0.

Up to 200°C, removal of water does not occur (Amandosov *et al* 1981). The loss of 0.1 wt % was recorded at $\simeq 220$°C, the dehydration rate reaching a maximum at 280°C (Gallagher 1976). According to Zhigarnovskii *et al* (1984), the loss of 1 wt% over 2 hours was observed at 212°C. The difference between these data seems to be related to the humidity of the medium in which the experiments were performed.

The vapour equilibrium pressure over KH_2PO_4 crystals was measured by Kiehl and Wallance (1927). The author and co-workers repeated these measurements using crystals with a deuteration degree of 0% and 95%. The results are brought together in table 1.4. They are described quite well by the Clapeyron equation for an ideal gas

$$\ln p = A - B/T$$

where $B = Q/R$, Q is the evaporation heat (more exactly, the activation energy of salt decomposition), R is the gas constant, and T is the absolute temperature. The approximation error, σ_0, and the equation constants are also given in table 1.4.

The analysis of the data represented in table 1.4 reveals a serious divergence for KH_2PO_4. It is to be noted that the results reported by Kiehl and Wallance (1927) contradict the fact that dehydration does not occur below 200°C. Indeed, the partial pressure of H_2O in the atmosphere at 25°C cannot exceed 22 mm Hg (100% humidity), therefore at 180°C (see table 1.4) the crystal should become dehydrated even in such a humid medium. At the same time a significant difference in p and Q for KH_2PO_4 and KD_2PO_4 is clearly evident. It is difficult to explain this difference, especially taking into account that heats of evaporation for H_2O and D_2O are very close ($\simeq 8\,\text{kcal mol}^{-1}$ at 200°C).

1.2.3 Solubility isotherms

The 25°C and 50°C isotherms of the $K_2O-P_2O_5-H_2O$ system were studied by Berg (1938a,b) and the 0°C isotherm was investigated by Ravich (1938).

Table 1.4 Water vapour pressure, p (mmHg), over $K(D_xH_{1-x})_2PO_4$ crystals at different temperatures, t(°C).

	$x = 0$				$x = 0.95$	
Data of Kiehl and Wallance (1927)			The author's data			
t	p	t	p	t	p	
180	25.5	194	6.6	145	2.5	
191	46.5	207	30.1	174	21	
210.5	118.7	217	100.6	198	71.2	
230	306.5	221	151.3	202	84.7	
250	673.3			226	228.2	
264	751					
A	28.2 ± 1.2	59.9 ± 1.2		29.9 ± 0.8		
B (K^{-1})	$11\,400 \pm 615$	$27\,100 \pm 560$		$12\,100 \pm 380$		
σ_0	0.28	0.05		0.11		
Q (kcal mol^{-1})	22.6 ± 1.2	53.8 ± 1.2		24.0 ± 0.8		

The 25°C isotherm of the $K_2O-P_2O_5-D_2O$ system (the deuteration degree of solutions being more than 95%) was studied by Barkova and Lepeshkov (1968). All those results are given in figures 1.9 and 1.10. The solubility isotherm branches of $K(H,D)_2PO_4$ as well as $NH_4H_2PO_4$ are actually linear (for the constants of the equations see table 1.16). The solubility

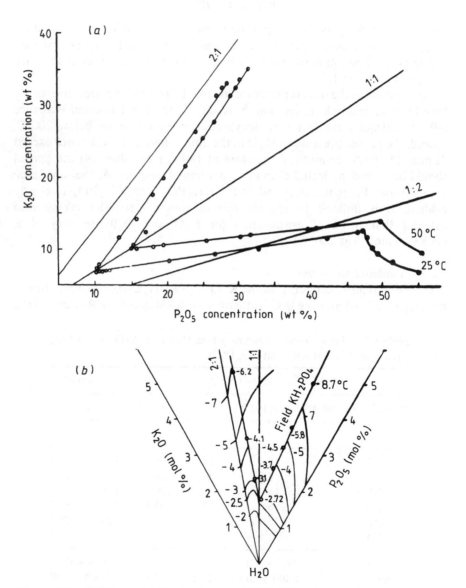

Figure 1.9   (*a*) The solubility isotherms of KH_2PO_4 (Berg 1938a,b); and (*b*) the crystallization isotherms of ice (Kurnakov *et al* 1938).

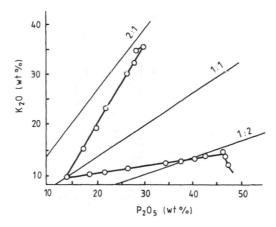

Figure 1.10 The solubility isotherms of KD_2PO_4 in 24°C (Barkova and Lepeshkov 1968).

minimum at all temperatures is very sharp and corresponds to a solution of stoichiometric composition. The solubility minimum singularity of KH_2PO_4 was specially verified by Krasil'shchikov (1933). He found that at 0°C the solubility curve had a singularity point with two tangents, but when the temperature was increased to 50°C the solubility minimum became less sharp and he suggested that the singularity point disappeared due to the increasing dissociation of the salt.

Figures 1.9 and 1.10(a) also show that the ultra-acid salt dissolves in water incongruently as in the system described previously (figure 1.1(a)). The ice crystallization field (figure 1.9(b)) as in the above system, is characterized by a kink in the isotherms on the line referring to the molar concentration relation of K_2O and P_2O_5 equal to unity.

It should be pointed out that both branches of the solubility isotherms do not exhibit any peculiarities indicating the presence of two solid modifications of the mono-substituted potassium orthophosphate. This fact appears to be due to both phases having similar solubilities.

There are a substantial number of studies of the solubility of KH_2PO_4 as a function of the solution pH. A graph from the work by Punin *et al* (1975) is shown in figure 1.11. The pH values are given for the solution saturated at 25°C.

1.2.4 Properties of solutions

(a) *Density*
Mullin and Amatavivadhana (1967) determined the density of aqueous solutions of KH_2PO_4 with the concentration ranging from 4.8 wt % to

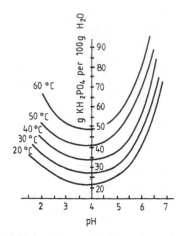

Figure 1.11 The effect of pH on the solubility of KH_2PO_4 in the $K_2O-P_2O_5-H_2O$ system (Punin *et al* 1975).

23 wt % at 20°C, 30°C and 40°C. We approximated their results by the following linear equations:

$$20°C: d = (0.992 \pm 0.001) + (0.00787 \pm 0.00009)c \pm 0.001$$

$$30°C: d = (0.989 \pm 0.002) + (0.00780 \pm 0.00011)c \pm 0.002 \quad (1.12)$$

$$40°C: d = (0.985 \pm 0.002) + (0.00783 \pm 0.00011)c \pm 0.002.$$

Using the data on solubility obtained by Mullin and Amatavivadhana for the saturated solution density in the same temperature range gives

$$d = (1.080 \pm 0.006) + (0.00263 \pm 0.00013)t \pm 0.003. \quad (1.13)$$

As in the case of $NH_4H_2PO_4$, equation (1.12) can be transformed to the form of (1.5), the mean-square approximation error being the same (for 20°C) or even less (0.0006 for 30°C and 40°C instead of 0.002). The corresponding partial densities are given in table 1.5.

According to Fedotova and Tsekhanskaja (1984), d_w and d_s depend appreciably on the solution concentration. Their results are given in figure 1.12 in the scale of molar volumes $v = M/d$, where M is the molar weight.

The density of saturated solutions of KD_2PO_4 in D_2O, according to Barkova and Lepeshkov (1966), is

$$d = 1.24 \pm 0.00315t \pm 0.007. \quad (1.14)$$

Krasil'shchikov (1933) determined the density isotherms of saturated solutions at 0°C and 25°C over a wide range of concentrations for K_2O and P_2O_5. As with the corresponding solubility isotherms, the curves for the density dependence on the solution composition have a sharp minimum referring to solutions of KH_2PO_4 in pure water (figure 1.13). Fifty years later

Table 1.5 The partial densities of water and KH_2PO_4 in solution ($g\,cm^{-3}$) by Mullin and Amatavivadhana (1967).

$t(^\circ C)$	d_w	d_s	B	σ_0
20	1.000 ± 0.006	2.85 ± 0.03	0.6490 ± 0.0010	0.001
30	0.997 ± 0.004	2.81 ± 0.02	0.6453 ± 0.0006	0.006
40	0.993 ± 0.004	2.83 ± 0.02	0.6496 ± 0.0007	0.007

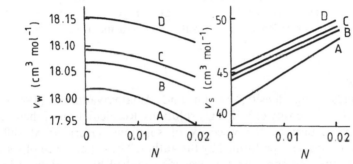

Figure 1.12 Concentration dependence (on the mol fraction scale) of the partial molar volumes of water v_w and dissolved KH_2PO_4 v_s at different temperatures, t ($^\circ C$): A, 10; B, 25; C, 30; D, 40.

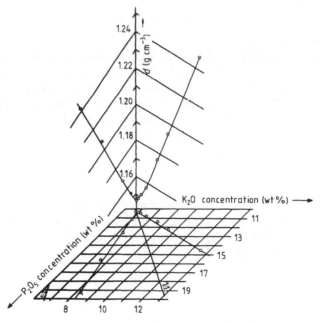

Figure 1.13 Density of KH_2PO_4 saturated solutions at $25^\circ C$ (Krasil'shchikov 1933).

similar measurements were performed at 25°C by Luff and Reed (1983). They approximated their results by a polynomial

$$d = 1.008 + 6.761 \times 10^{-3}P + 9.439 \times 10^{-3}K + 5.67 \times 10^{-5}P^2$$
$$+ 12.07 \times 10^{-5}K^2 + 6.13 \times 10^{-5}PK \pm 0.004$$

where P and K are the P_2O_5 and K_2O concentrations in wt %, respectively. The authors did not report the experimental data used to derive the polynomial. Differentiating the expression for d it is found that the density minimum corresponds to the molar ratio $K_2O/P_2O_5 = 0.94$. This is consistent with Krasil'shchikov's results which appear to be much more reliable.

(b) *Viscosity*
Besides measuring density, Mullin and Amatavivadhana (1967) also determined the viscosity of KH_2PO_4 solutions. Recalculation of their results in order to determine the viscosity of solutions saturated at different temperatures has been performed by the author. The dependence of viscosity on temperature and concentration was also studied by Sokolowski (1981). The viscosity of saturated KD_2PO_4 solutions was measured by Barkova and Lepeshkov (1966). All these results are summarized in figure 1.14(a). Figure 1.14(b) shows the isoviscosis in a ternary system examined by Ravich (1940) at 0°C.

(c) *Diffusion coefficient*
The diffusion coefficient of KH_2PO_4 between two solutions was measured by Mullin and Amatavivadhana (1967) in the temperature range 15–45°C and by Rosmalen (1977). The temperatures of the solutions differed by 5°C, each being 0.5°C higher than that of saturation. In the temperature range mentioned above, the diffusion coefficient increased from 5.5×10^{-6} to $9.4 \times 10^{-6} \, cm^2 \, s^{-1}$. The product of viscosity and the diffusion coefficient increases linearly with temperature.

(d) *Conductivity*
Using the standard technique involving platinum electrodes, Sokolowski and Kibalczyc (1981) determined the concentration dependence ($c > 3$ wt %) of the conductivity for KH_2PO_4 solutions over the temperature range 30–38°C. According to their data,

$$\chi = \chi_0[1 + \alpha(t - t_0)]$$

where $\alpha = 0.0173 \, (\pm 0.0004)$, $t_0 = 30.54$°C, and χ_0 is the specific conductivity at $t = t_0$. At $t = t_0$

$$\chi_0 = A \ln c + B$$

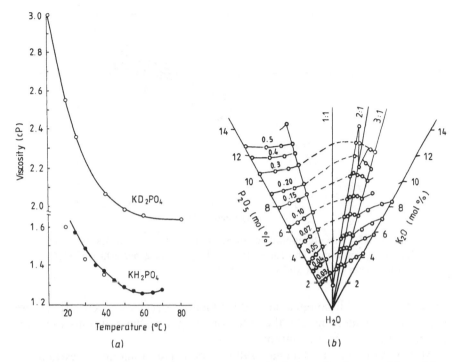

Figure 1.14 (a) Viscosity of KD_2PO_4 saturated solutions (Barkova and Lepeshkov 1966) and that of KH_2PO_4: $\bigcirc$, Mullin and Amatavivadhana (1967); $\bullet$, Sokolowski (1981). (b) Isoviscosity lines in the $K_2O-P_2O_5-H_2O$ system at $0°C$ (Ravich 1940). The viscosity in poise is given at the curves.

where $A = 0.03522\,\Omega^{-1}cm^{-1}$, $B = -0.01364\,\Omega^{-1}cm^{-1}$, the approximation error being $\pm 0.00009\,\Omega^{-1}cm^{-1}$.

Bordui and Loiacono (1984) measured conductivity with a transformer transducer, which was first used for such measurements by Agbalyan et al (1975). This transducer produces no space charges in solution which makes measurements much easier. The experiments were carried out at $31-41°C$ in solutions of 11 different concentrations. The results are given in figure 1.15.

Using this technique together with computer processing, the authors estimate the concentration determination accuracy, according to data on χ and t, to be ± 0.05 g per 100 g of H_2O ($\pm 0.15\%$ of the relative supersaturated solution).

Bordui et al (1985) employed the transducers that controlled variables χ and t to automate crystal growth by maintaining a constant supersaturation of the solution in the production process (to an accuracy of $\pm 0.3\%$). They were the first to effect continuous control of supersaturation.

The results obtained by Bordui and Loiacono (1984) differ from those by Sokolowski and Kibalczyc (1981): from figure 1.15 we obtain $\alpha \simeq 0.0200$

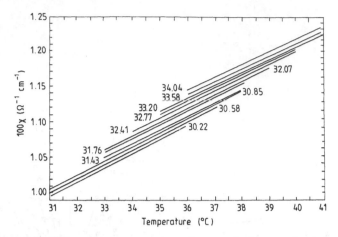

Figure 1.15   The temperature dependence of the conductivity of KH_2PO_4 solutions. The concentration is given in g per 100 g of H_2O at the curves (Bordui and Loiacono 1984).

instead of 0.0173, although a linear dependence of $\chi(t)$ also occurred in these experiments. Unfortunately, the value of the pH of the solutions used has not been reported by the authors.

Shimomura and Tachibana (1986) conducted analogous research at 10–70°C with solutions of four different concentrations (pH in the range 3.95–4.15). In the curves of $\chi(1/T)$ they observed a kink at the saturation temperature. In other works such a kink has never been observed.

(e) Surface energy
The surface energy of the crystal–saturated solution interface has generally been determined by the dependence of the nucleating rate upon saturation. The values of α varied from 1.9 erg cm^{-2} (Wojciechowski and Kibalczyc 1986) to 16 erg cm^{-2} (Bennema et al 1973). Shanmugham et al (1984, 1985) obtained $\alpha = 5.9$ erg cm^{-2}. Using dissolution kinetics data, Koziejowska and Sangwal (1988) obtained 3.6 erg cm^{-2} for {011} faces and 2.8 erg cm^{-2} for (010) (at 20°C). Calculations by Söhnel (1982) on the basis of the crystallization heat data (with $\Delta H = 4.1$ kcal mol^{-1}) gave $\alpha = 24$ erg cm^{-2}. Different values of α are treated in more detail in Chapter 2.

(f) Heat capacity
The heat capacity of solutions in the K_2O–P_2O_5–H_2O systems were determined by Luff and Reed (1983) at 25.7 ± 0.4°C. They approximated their results by the following polynomial

$$s = 0.9847 - 8.699 \times 10^{-3}P - 12.67 \times 10^{-3}K - 2.87 \times 10^{-5}P^2$$

$$+ 1.53 \times 10^{-5}K^2 + 21.59 \times 10^{-5}PK \pm 0.003 \text{ cal g}^{-1} \text{ deg}^{-1}$$

where P and K are concentrations of P_2O_5 and K_2O, respectively, in wt %. For the solution of stoichiometric composition, saturated at this temperature, we obtain $s = 0.818$ cal g^{-1} deg^{-1}; the heat capacity of water at 25.7°C being $s = 0.985$ cal g^{-1} deg^{-1}.

(g) *Refraction index*
The refraction index of KH_2PO_4 solutions as a function of temperature and concentration was determined by Krasinski *et al* (1984). Recalculation of these data for the refraction index of saturated solutions as a function of the saturation temperature can be expressed as follows

$$n = (1.3473 \pm 0.0004) + (3 \pm 0.1) \times 10^{-4}t \pm 0.0003.$$

The results obtained by Takubo and Makita (1989) are shown in figure 1.16. The dependence $n(c)$ at 40°C has the form

$$n = 1.3306 + 0.001\,1645c.$$

For saturated solutions (in the range 10–60°C)

$$n = 1.3486 + 0.0002858t.$$

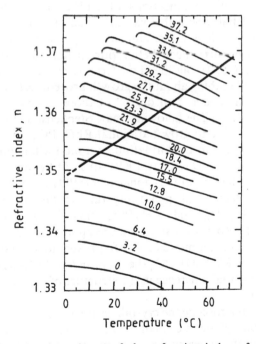

Figure 1.16 The dependence of the refractive index of a KH_2PO_4 solution on temperature and concentration (Takubo and Makita 1989). The straight line represents compositions of saturated solutions.

1.3 The $Rb_2O-P_2O_5-(H,D)_2O$ system

The rubidium system has been studied less extensively. It differs from the systems described in sections 1.1 and 1.2 because the process of phase formation in deuterated solutions is more complex. In this case three crystallographically different mono-substituted orthophosphates are produced and the conditions of growth of both monoclinic crystals with $x = 0$ and tetragonal ones with $x = 1$ are still not very well known. Plyushchev and Stepin summarized the related data available up to 1968 in their book published in 1970.

1.3.1 Solid phases

Berg was the first to synthesize RbH_2PO_4, as well as CsH_2PO_4, in 1901. The synthesis was repeated in 1927 (Hendricks 1927). Berg reported (Berg 1901) that besides tetragonal RbH_2PO_4 crystals he also observed biaxial (presumably monoclinic) crystals. Hendricks succeeded in obtaining only biaxial crystals, although he did not study them, because it was the structure of tetragonal phases isomorphous to KH_2PO_4 that interested him. The ferroelectric properties of tetragonal RbH_2PO_4 were discovered in 1945 (Bärtschi et al 1945, Matthias et al 1947). It should be noted that Bärtschi and co-workers studied the temperature dependence of the dielectric susceptibility for both modifications of RbH_2PO_4. Unfortunately, they did not provide details about the way they had grown the biaxial crystals. These days, however, it is rather difficult to grow monoclinic RbH_2PO_4 crystals, perhaps because much better purification of starting material is required. Matthias (1952) considered the ferroelectric properties of tetragonal RbD_2PO_4 but said nothing about the growth procedure. (Twenty years later we showed that the deuteration degree of those samples was about 78% (Volkova et al 1971a).) The following year, Stephenson et al (1953) argued that it was not possible to grow tetragonal RbD_2PO_4 crystals with a high deuteration degree due to the formation of a monoclinic phase. This point of view still existed in 1969 (Adhav 1969). In 1970 Remoissenet et al briefly mentioned the production of tetragonal RbD_2PO_4 crystals.

The habit of tetragonal phase crystals is analogous to those of tetragonal $NH_4H_2PO_4$ and KH_2PO_4 crystals. The idea that PO_4 tetrahedrons in the RbH_2PO_4 structure are appreciably distorted (Rez et al 1967, Pakhomov and Sil'nitskaya 1970) proved to be wrong. Just as in $K(H,D)_2PO_4$, so in $Rb(H,D)_2PO_4$ the shape of tetrahedrons is nearly perfect and there is no essential difference in the structure of these crystals (Rusakov et al 1978, 1979).

Viewing with a microscope the crystallization process of mono-substituted rubidium orthophosphate in a small volume of stoichiometric solution with different deuteration degrees in the temperature range 5–50°C, Mishchenko and Rashkovich (1971) found that crystals of differing symmetry and habit were produced depending on the experimental conditions. The possibility of seeding the liquid phase allowed the author to determine quite accurately the existence regions of separate solid phases. Subsequently the corresponding crystals were grown and described, their solubility was determined and the effect of the solution acidity on the position of the existence regions was ascertained (Rashkovich and Momtaz 1978).

We called rhombic plates, crystallizing from a solution with $x \gtrsim 0.6$ under appropriate temperatures (see below), phase I (figure 1.17).

The crystals have the point symmetry group $2/m$. There external habit is represented by the faces of a pinacoid $\{100\}$ (they are the most developed) and two rhombic prisms $\{011\}$ and $\{112\}$. Goniometric measurements yielded $a:b:c = 0.81:1:0.73$, with the error being $\pm 0.8\%$, and $\beta = 107°30' \pm 30'$. Reflections with indices $(h00)$ were taken from pinacoid faces using an x-ray diffractometer and parameter $a = (7.61 \pm 0.01)$ Å was determined. The density of the phase I crystals measured by hydrostatic weighing in toluene was $2.845 \pm 0.005\,\mathrm{g\,cm^{-3}}$.

Upon heating, phase I decomposes at the same temperature as tetragonal crystals ($\simeq 230°C$). Below this temperature no weight loss is observed, in decomposition $\simeq 10.5\,\mathrm{wt\%}$ is lost, which corresponds to the loss of one water molecule. Thus, the composition of this phase is identical to that of tetragonal crystals.

In air the phase I crystals are coated with a white film (with $x > 0.85$) or they crack (with $x < 0.8$). The x-ray analysis showed that at the same time, the transition into the tetragonal phase occurred. The transition time depends on the size and deuteration degree of a given crystal. Thus, $5 \times 5 \times 1$ mm crystals with $x = 0.95$ persist for several years, while those with $x = 0.8$ become opaque and crack in less than a week after their removal from solution. In powder, the transformation already occurs to a large measure during the grinding of the crystal; a Debyegram taken at this stage detects the tetragonal phase peaks. However, the tetragonal phase does not appear even in powder with $x \simeq 0.99$. These crystals exhibit the piezoeffect several days after their removal from the solution, which results from the formation of a non-centrosymmetrical tetragonal phase on the sample surface.

Figure 1.17   Phase I crystals grown from solution with $x = 0.8$ at $50°C$, after their transition into the tetragonal phase.

Long thin plates (approximate dimensions are 200:50:1) crystallized with $x < 0.7$ at relatively high temperatures. Such crystals grew at high supersaturations in solutions with $x = 0$ (for example, if a solution saturated at 50°C is cooled actively with cold water). These plates cracked 2–3 hours after removal from solution intact, but if crystallized from H_2O, they broke up into powder. In the latter case, if left in solution they dissolve completely over about one day and recrystallize into tetragonal crystals. Starting from the fact that they behave in deuterated solutions just as the rhombic crystals described above, we believe that this is phase I, with the crystal habit having been changed due to the high supersaturation of solutions.

Isometric crystals, that were called phase II (figure 1.18), precipitate from solutions with $0.4 < x < 1$ at relatively low temperatures (less than 36–40°C). These crystals also refer to monoclinic symmetry (point group $2/m$). The habit of the crystal shown in figure 1.18 is a combination of pinocoids $\{100\}$, $\{001\}$ and a rhombic prism $\{011\}$. In crystals with $x > 0.8$, grown at 36–40°C, a pinocoid $\{010\}$ is also observed, but faces $\{001\}$ are grown more intensively joining both faces $\{100\}$ (figure 1.19). According to goniometric measurement $a:b:c = 1.12:1:0.85$, $\beta = 104°30' \pm 30'$, the density of crystals, with $x > 0.85$, is $d = 2.65 \pm 0.01\ \mathrm{g\,cm}^{-3}$.

Upon heating, phase II decomposes at 40–41°C releasing one water molecule and transforming into the tetragonal phase which behaves normally upon further heating. Thus, this modification of mono-substituted rubidium orthophosphate contains one molecule of crystal water. The loss of water occurs faster in air at room temperature, if the humidity is lower. Upon grinding the crystals, the powder becomes moist, i.e. grinding speeds up decomposition. Therefore Debyegrams taken 20 minutes after grinding detect the tetragonal phase peaks only.

Figure 1.18   Phase II crystals grown from solution with $x = 0.8$ at 30°C.

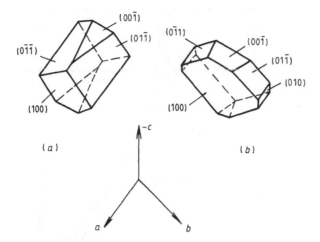

Figure 1.19 The form of phase II crystals grown with (*a*) $x < 0.8$, and (*b*) $x > 0.8$.

There is one important point to be mentioned as far as phase II crystallization is concerned. In extremely pure solutions, as well as in ordinary solutions preheated to temperatures higher than 40°C and allowed to stand for some time, spontaneous (i.e. with no seeding) crystallization of phase II occurs only upon intense supercooling (to 10°C or less). It appears that certain uncontrolled impurities promote the stabilization of phase II.

Figure 1.20 shows the spontaneous crystallization fields of the phases described above, the field being separated by the coexistence curves of the two phases. Metastable parts of these curves are denoted by the broken lines. Plotting the curve COD involves considerable experimental difficulties because of the intense evaporation of the solution upon seeding, therefore the curve plotted is not quite accurate.

The section AO of the curve AOB determines the crystallization regions of phase II and the tetragonal phase. In solutions with $x < 0.4$, formation of the phase II crystals has never been observed. Phase II seeding crystals with $x \simeq 0.5$ dissolved when introduced into a solution with the same deuteration degree. In the section OB, phase II coexists with phase I. Here the tetragonal seed dissolves in the presence of phase II. If there are no phase II crystals, the tetragonal seed growth is not observed, with phase I crystallizing on its surface. Thus, the tetragonal phase with $x > 0.7$–0.75 cannot crystallize spontaneously at any solution temperature, with 40°C being the upper temperature limit for completely deuterated crystals to exist in solution.

The curve COD is plotted on the basis of the following data. Observations of crystallization from solutions with $x = 0$ at 65–90°C showed that above $\simeq 75$°C, with the solvent evaporating, very small crystals are crystallized

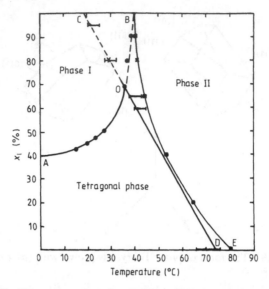

Figure 1.20   The existence regions of the solid phases of the $Rb(H,D)_2$ $PO_4-(H,D)_2O$ system. Curve BE corresponds to the onset of the transition of the tetragonal phase into the monoclinic one upon heating in air (D'yakov *et al* 1973). The intersections of the curves given in figure 1.21 are marked by crosses.

from the solution. X-ray diffraction showed their structure to be different from that of the tetragonal form. These crystals are not stable in air at room temperature, rapidly transforming into the tetragonal phase. Above 70°C the tetragonal RbH_2PO_4 seed plates become oblique and crack. Increasing the degree of deuteration results in lowering the upper temperature limit of the tetragonal phase stability. With $x = 0.6$, the limiting temperature is 45°C; it is phase I crystals that grow at higher temperature. In the experiments with $x = 0.65$ the tetragonal phase crystals in a supersaturated solution ($\simeq 40$°C) are covered by phase I crystals grown around them. The same situation was observed in solution with $x = 0.8$ at 28–30°C, when phase II crystals were absent. In the triangular COB region it is only phase II that crystallizes spontaneously, but if it is absent, then phase I crystals can grow and exist in this region. The curve OC is the coexistence curve of phase I and the tetragonal phase, with phase II crystals being absent. Thus, if in solution with $x > 0.9$ a tetragonal crystal seed plate is placed under a layer of phase I crystals at a temperature corresponding to the curve OC and the temperature is then lowered to 5–10°C, then a tetragonal crystal grows. These data are represented in figure 1.20 as line segments encompassing the marked temperature regions.

Tetragonal crystals can also be grown in the phase II crystallization region by crystallizing on seeds from solutions with $x < 0.7$ provided the temperature is lowered slowly, thereby avoiding the high supersaturation of solution necessary for spontaneous phase II crystallization.

Two remarks should be made concerning the phase rule in connection with figure 1.20. It is known that two phases of the same composition can coexist only in a one-component system. Therefore the tetragonal phase and phase I of different deuteration degree will be crystallized from solutions with a deuteration degree and temperature corresponding to the curve OD. The deuterium distribution coefficient between solution and phase I crystals could not be determined because of their instability. However, in the analogous situation with $K(H,D)_2PO_4$ the distribution coefficients for tetragonal and monoclinic crystals were indeed different (see below). Then, it should be emphasized that the system in question (a solution of mono-substituted rubidium orthophosphate) is not actually a two-component system, but a three-component one, because the deuterium content varies. Therefore the presence of the invariant point O, where the solution and the three solid phases coexist, does not contradict the phase rule.

From the above considerations it follows that in solutions with $x = 0.9-1.0$ at temperatures in the range 20–25°C, phase II should have the lowest solubility, while phase I should have the highest. The solubility of the tetragonal phase occupies the intermediate position. It is shown below that an experimental determination of the solubility curves confirms the arguments presented. All these facts being taken into account, a technique for growing tetragonal crystals of mono-substituted rubidium orthophosphate with a high deuteration degree has been proposed. The idea behind the technique is that the supersaturation required for growing tetragonal crystals is produced by the more soluble phase I crystals. The main difficulty here is to prevent crystallization of phase II. The technique is as follows. A saturated solution is prepared in the region of phase I spontaneous crystallization, i.e. at temperatures higher than 40 45°C. Upon cooling, phase I crystals precipitate at the bottom of the crystallizer. If cooling is performed slowly and without stirring, the solution can be cooled to 20°C and beyond to lower temperatures (i.e. the system can be transferred to the left of curve CO) without phase II being crystallized. If now the solution is inoculated with the tetragonal phase, the seed will grow at the expense of dissolving phase I. The necessary growth rate is controlled by the experimental temperature, the higher the temperature, the less is the difference in solubilities of both phases and the less is the solution supersaturation with respect to the tetragonal phase. The disadvantage of this technique is that phase II can often spontaneously crystallize from the solution, with the other two phases dissolving quickly.

Numerous experiments have been carried out to study the effect of the solution acidity on the formation of highly deuterated tetragonal phase crystals. The position of the curves in figure 1.20 was found not to depend appreciably on the content of acid and alkali in the solution.

Sumita et al (1981) produced monoclinic RbD_2PO_4 crystals (phase I) by evaporating a solution of RbH_2PO_4 in D_2O and subsequent recrystallization in D_2O. In these crystals they found prominent cleavage along the planes (010). They grew phase I crystals with $x = 0.5$ and 0.7, and they believe that $x = 0.2-0.3$ is the low limit for the existence of this phase. However, Sumita and co-workers say nothing about phase II. Komukae and Makita (1985) obtained phase I with $x = 0.3$.

Averbuch-Pouchot and Durif (1985) described monoclinic RbH_2PO_4 crystals (point group P2/m, axes ratio being 0.806:1:0.649) obtained by the reaction $Rb_4P_4O_{12} + 4H_2O \Rightarrow 4RbH_2PO_4$. The solution was heated at 60°C until it was transformed into a gel which was kept at room temperature in a moist medium. Tetragonal crystals were also grown by Sumita et al (1981) with RbH_2PO_4 seeds from a supersaturated solution of RbH_2PO_4 in D_2O. Some properties of RbH_2PO_4 monoclinic crystals were studied by Baranowski et al (1986, 1987), those of RbD_2PO_4 were examined by Baranov et al (1987).

1.3.2 Solubility polytherms

The solubility of RbH_2PO_4 in H_2O at 25°C was determined by Zvorykin and Vetkina (1961). The temperature dependence of the solubility at 0–80°C was studied by Bykova et al (1968). These data are given in table 1.6. The accuracy of the concentration determination is not specified in the work of Bykova et al (1968), the solution stoichiometry was also not determined.

In experiments by the author and co-workers on solubility determination, the solution was prepared from RbH_2PO_4 single crystals. The concentration of the solution after reaching equilibrium was determined in two ways: by slow drying the solution sample to constant weight (the temperature was

Table 1.6 Solubility of RbH_2PO_4 in H_2O (wt %).

Temperature (°C)	The author's data	Data by Bykova et al (1981)	Temperature (°C)	The author's data	Data by Bykova et al (1981)
0	—	30.2	35.0	47.9	—
8.8	36.8	—	35.5	48.8[a]	—
14.0	39.0	—	40.0	—	50.9
16.0	39.6	—	42.0	51.3	—
21.8	41.9	—	45.0	52.5	—
22.8	42.8[a]	—	50.0	54.7	55.3
25.0	43.5	44.0	52.1	55.6	—
26.2	44.2	—	60.0	—	57.8
30.0	45.0	—	80.0	—	62.0

[a] Determined by the concentration flow method.

slowly raised to 110°C, the process lasting for 10 days); and by calcinating the solution sample at 800–900°C. The accuracy of the concentration determination was 0.1 wt %. Moreover, the saturation temperature was determined by the concentration flow method. The results are given in table 1.6. For the temperature range 8–50°C, over which the experiments were carried out, the linear approximation has the form

$$c = (32.68 \pm 0.11) + (0.440 \pm 0.003)t \pm 0.17. \qquad (1.15)$$

Processing the data of Bykova *et al* (1981) gives the following relation

$$c = 30.67 + 0.502t \pm 0.7.$$

In a similar way, we determined the solubility of three $Rb(H,D)_2PO_4$ modifications in the solution with a deuteration degree of 80%. It is difficult to determine the tetragonal phase solubility with a higher deuteration degree, since under these conditions where intense crystallization occurs one cannot be sure that no crystals of other solid phases are present in the system.

The experimental results are presented in table 1.7 and figure 1.21. The experimental data can be approximated by the following expressions

$$C_{tetr} = (37.6 \pm 0.3) + (0.507 \pm 0.014)t \pm 0.1 \qquad (1.16)$$

$$C_{phII} = (34.9 \pm 0.7) + (0.514 \pm 0.034)t \pm 0.7 \qquad (1.17)$$

$$C_{phI} = (43.7 \pm 0.4) + (0.300 \pm 0.010)t \pm 0.1. \qquad (1.18)$$

One can see from figure 1.21 that the solubility curves for tetragonal and phase II crystals do not intersect; this should be expected from the analysis given above of the data presented in figure 1.20. The solubility curves for tetragonal and phase II crystals do intersect with the phase I solubility curve at 29°C and $\simeq 41$°C, respectively. These points are marked with crosses in figure 1.20; they are seen to correspond to the results of observations described above.

Table 1.7 Solubility of $Rb(H,D)_2PO_4$ in solution with $x = 0.80$ (wt %).

Temperature (°C)	Tetragonal phase	Temperature (°C)	Phase II[a]	Temperature (°C)	Phase I
17.2	45.4	2.0	35.9	35.0	54.2
22.3	48.8	20.9	45.0	39.0	55.3
23.0	49.6	24.0	48.0	41.0	56.1
25.0	50.2	30.4	50.3	42.0	56.4
26.2	51.0	—	—	45.0	57.2
—	—	—	—	50.0	58.6

[a] The concentration is calculated for the anhydrous salt.

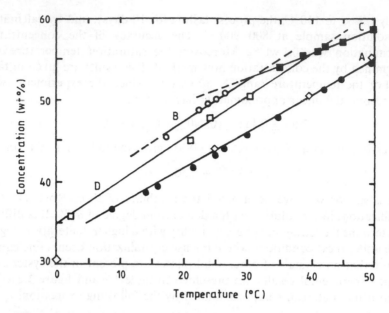

Figure 1.21 $Rb(H,D)_2PO_4$ solubility in H_2O (A) and in solution with $x = 0.8$: tetragonal phase (B), phase í (C) and phase II (D). On line A, $\diamond$ represent the data of Bykova *et al* (1981).

Melting RbH_2PO_4 in a closed container takes place at 283°C (Zhigarnovskii *et al* 1984). In air the release of water occurs in two stages: 0.25 mol H_2O being released between 185 and 250°C, the remaining 0.75 mol H_2O being released between 250 and 420°C. Among the decomposition products the double salt $2RbH_2PO_4 \cdot Rb_2H_2P_2O_7$ and the pyrophosphate $Rb_2H_2P_2O_7$ are fixed (Nirsha *et al* 1981, Gallagher 1976). In contrast to KH_2PO_4, dehydrated rubidium orthophosphate does not dissolve readily in water and forms a gel.

1.3.3   Solubility isotherms
Solubility isotherms at 25°C and 50°C were studied by Rashkovich and Momtaz (1978). Solutions in contact with a solid phase were kept at a constant temperature and were stirred periodically for at least two weeks, although only 5 days are required for equilibrium to be reached. Two samples (2 g each) were taken from each solution. Immediately after weighing, the samples were diluted with H_2O in the ratio 1:50 (by weight). Following this, potentiometric titration of KOH was used to determine the solution composition. Alkaline solutions were acidified with a certain amount of H_3PO_4. The relative error in determining the concentration of Rb_2O and P_2O_5 was $\pm$ 0.2%.

In the experiments at 25°C, the pH of saturated solutions was also determined by using a standard microdish for a pH-meter.

In non-deuterated solutions the tetragonal phase crystals were in equilibrium with the solution.

In the experiments with D_2O, solutions were prepared in such a way that after reaching equilibrium their deuteration degree was 90–91%. At 25°C, phase II crystallized from solutions of any acidity, at 50°C phase I crystallized.

The experimental data on the system with $x = 0$ are given in table 1.8, those on the system with $x = 0.9$ are given in table 1.9. All four isotherms are shown in figure 1.22. Just as in the case of the systems described in the two previous sections, the acid and alkaline branches of the isotherms are nearly straight lines. Constants of the isotherm linear equations are given in table 1.16 (see later).

The solubility of RbH_2PO_4 in H_2O (determined by interpolating the data given in table 1.8 and figure 1.22) was 44.1 wt % at 25°C and 55.2 wt % at 50°C, which is in good agreement with the data given in table 1.6.

In stoichiometric solutions with $x = 0.9$ the solubility of phase II at 25°C was 51.7 wt % and that of phase I at 50°C was 58.1 wt %. These results differ slightly from those in table 1.7, which is associated with changing the degree of deuteration. Positions of invariant points were not determined, but in some experiments the ultra-acid salt $(Rb(H,D)_2PO_4.(H,D)_3PO_4)$ was in equilibrium with a solution of low pH. This is clearly seen by the altered habit of the large-sized crystalline precipitate. To determine the solid

Table 1.8 Composition of solutions in equilibrium with RbH_2PO_4 at 25°C and 50°C (wt %). pH refers only to 25°C.

pH	25°C		50°C	
	Rb_2O	P_2O_5	Rb_2O	P_2O_5
6.6	35.4	19.9	36.2	21.8
5.8	29.1	18.3	33.5	21.5
5.6	27.6	17.9	31.8	21.1
5.2	26.7	17.4	31.3	21.3
5.2	25.3	17.4	30.1	21.2
4.7	23.8	17.2	28.6	20.8
2.6	24.6	23.8	28.3	21.2
2.5	24.9	24.6	28.5	23.2
2.1	25.8	27.5	29.5	28.4
1.9	26.6	30.2	30.3	33.6
2.0	27.4	32.2	30.7	36.0
1.4	28.3	36.3	31.3	39.9
			31.4	39.5
			30.9[a]	41.7[a]

[a] Ultra-acid salt is in equilibrium with solution.

Table 1.9 Composition of solutions with $x = 0.9$ in equilibrium with solid phases at 25°C and 50°C (wt %).

25°C; solid phase II		50°C; solid phase I	
Rb_2O	P_2O_5	Rb_2O	P_2O_5
43.2	22.2	42.9	23.2
35.5	21.2	36.5	22.5
32.3	20.3	34.5	22.4
31.0	20.1	33.5	22.3
28.8	19.9	32.0	22.3
26.4	20.8	29.9	22.5
28.2	29.2	29.6	23.2
28.9	32.6	30.5	30.2
28.8	32.9	30.8	33.1
29.4	35.9	31.1	33.4
26.1[a]	38.2[a]	31.5	36.5
		31.1[a]	37.8[a]

[a] Ultra-acid salt is in equilibrium with solution.

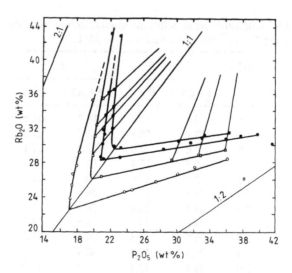

Figure 1.22 The solubility isotherms of $Rb(H,D)_2PO_4$ at 25°C (open symbols) and at 50°C (full symbols): circles, $x = 0$; squares, $x = 0.9$. The solution compositions plotted on the straight lines converging at the figuration point of $Rb(D_{0.9}H_{0.1})_2PO_4$ refer to experiments with the same solvent composition.

phase composition in such cases the following procedure was used. About 2 g of crystalline sediment was placed on filter paper to remove the remaining solution. Then the salt was dissolved in water and the solution obtained was titrated. The molar ratio of the Rb_2O and P_2O_5 concentrations was determined by the positions of the equivalence points (under these conditions it is not required to know the amount of the sample used). The analysis showed that $(Rb_2O/P_2O_5)_{mol}$ was slightly greater than 0.5 due to the incomplete removal of saturated solution by the filter paper. The positions of the corresponding points in figure 1.22 indicate that the ultra-acid salt dissolves congruently in water (in contrast to systems with potassium and ammonium).

Figure 1.22 shows the lines connecting equilibrium compositions of the liquid phase ($x = 0.9$) at 25°C and 50°C in the experiments with solvents of the same composition, i.e. in the experiments in which the system composition was not corrected after the concentration determination at 25°C. All these straight lines intersect in the narrow region of compositions near the figuration points referring to compositions of phase I and II. (The content of Rb_2O and P_2O_5 in these phases was equal to 50.7 and 38.5 wt %, and 45.7 and 34.7 wt %, respectively.) The figuration points of the solid phases are far removed from the equilibrium curves, therefore considering the accuracy of the experiments it is not possible to outline with certainty the intersection region of the straight lines under study.

An important feature of the solubility isotherms in the $Rb_2O-P_2O_5-(H,D)_2O$ system is a slight shift of the solubility minimum into the alkaline region from the line of solubility of stoichiometric solution compositions. Note that this minimum is much less sharp than in the systems discussed previously (figures 1.1, 1.9, 1.10).

It should also be noted that the isotherm behaviour in the systems with D_2O and H_2O is similar, even though it is solid phases of different structure that are in equilibrium with solution. This seems to indicate that there are the same complexes in solution regardless of the crystallizing solid phase structure, so that changing the solid phase structure is not connected with a qualitative change in the properties of the saturated solutions. In this connection, an investigation of the combination scattering spectra of the rubidium salt solutions was carried out (Mavrin *et al* 1973). Saturated solutions in H_2O and D_2O in contact with crystals of different phases (I, II and tetragonal) were studied; moreover, combinational scattering spectra of the solid phases mentioned above were taken. The anion symmetry was found to be independent of the anion deuteration degree and of the solid phase which was in contact with the saturated solution.

1.3.4 Properties of solutions

Unfortunately, the data given below for this aspect of the study are the only available ones, since these problems were studied solely by Mishchenko and Rashkovich (1973) thus the results cannot be compared with any others.

(a) *Density*

The density of solutions of constant concentrations at different temperatures was determined by weighing a quartz cylinder in the solution under study. The solution was placed in a thermostatic container with a magnetic stirrer. Measurements were performed upon cooling and heating the solution. When determining the saturated solution density, saturation was achieved by intense stirring of the solutions with finely dispersive solid phase for four hours. The concentration was determined by drying the solution sample to constant weight (to an accuracy of ± 0.1 wt %). The accuracy of the determination of the absolute density value was $\pm 0.05\%$ with a reproducibility of the experimental results of 0.01%. The temperature was stabilized to an accuracy of $\pm 0.05\%$. The experiments were carried out in solutions with $x = 0$ and 0.8.

The experimental data were obtained for nine concentrations (from 5 to $\simeq 55$ wt %) with $x = 0$, and for five concentrations (from 30 to 55 wt %) with $x = 0.8$. In all cases the density decreased linearly with increasing temperature, with the temperature coefficient increasing with increasing concentration (it is equal to 0.0003 g cm^{-3} °C^{-1} at 20 wt % and to 0.0006 g cm^{-3} °C^{-1} at 50 wt %). In deuterated solutions the temperature coefficient was about 15% greater. All experimental data were fitted to the empirical formula

$$d = a - a_1 t + (b - b_1 t)c^*$$

where c^* is the volume salt concentration: $c^* = cd/100$ g cm^{-3}. The values of the constants are given in table 1.10.

For example, in figure 1.23 the full curves show the density of solutions at constant temperature as a function of concentration. The curves are plotted according to the data of table 1.10 after transforming the reduced formula

Table 1.10   Constants of equations for the density of $Rb(H,D)_2PO_4$ solutions in water and relevant partial quantities.

	$x = 0$	$x = 0.8$
$a = d_{wo}$ (g cm^{-3})	1.0141 ± 0.0026	1.0987 ± 0.0054
a_1 (g cm^{-3} deg^{-1})	0.00050 ± 0.00011	-0.00078 ± 0.00021
b	0.7045 ± 0.0050	0.6739 ± 0.0079
b_1 (deg^{-1})	0.00051 ± 0.00014	0.00182 ± 0.00025
σ_0 (g cm^{-3})	0.0029	0.0020
Number of experiments	51	34
Temperature range (°C)	14–64	15.6–50.1
Concentration range (wt %)	5.0–56.7	32.9–55.6
α_w (deg^{-1})	0.0005 ± 0.0001	-0.0007 ± 0.0002
d_{so} (g cm^{-3})	3.43 ± 0.07	3.37 ± 0.10
α_s (deg^{-1})	0.00022 ± 0.00006	0.00049 ± 0.00011
d_{tetr} (g cm^{-3})	2.858	2.873

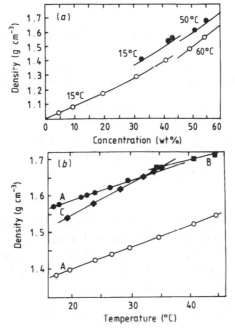

Figure 1.23 Density of Rb(H,D)$_2$PO$_4$ solutions: (*a*) concentration effect at a constant temperature (denoted at the curves); (*b*) saturated solutions: A, tetragonal phase; B, phase I; C, phase II. Open symbols, $x = 0$; full symbols, $x = 0.8$.

to the form

$$d = \frac{a - a_1 t}{1 - (b - b_1 t)c/100}.$$

The experimental points are also plotted there.

The same empirical formula can be written in the form

$$d = d_{wo}(1 - \alpha_w t) + \{1 - (d_{wo}/d_{so})[1 + (\alpha_s - \alpha_w)t]\}c^*$$

where α is the partial coefficient of thermal expansion, d_{wo} and d_{so} are the partial densities of water and salt, respectively. The values of the constants are also given in table 1.10 (where σ_0 is the root mean square of approximation). Data on the density of saturated solutions are shown in figure 1.23(*b*). One can see that the density depends linearly on the saturation temperature. Processing the data by the least-squares method gives:

$x = 0$,

$$d = (1.2831 \pm 0.0011) + (0.00570 \pm 0.00003)t \pm 0.0009 \qquad (1.19)$$

$x = 0.8$ (tetragonal phase),

$$d = (1.4765 \pm 0.0019) + (0.00533 \pm 0.00008)t \pm 0.0009 \qquad (1.20)$$

$x = 0.8$ (phase I),

$$d = (1.538 \pm 0.007) + (0.00388 \pm 0.00018)t \pm 0.0013 \qquad (1.21)$$

$x = 0.8$ (phase II),

$$d = (1.376 \pm 0.002) + (0.00844 \pm 0.00008)t \pm 0.0011. \qquad (1.22)$$

Equations (1.20) and (1.22) were obtained for the temperature region below 35°C, whereas (1.21) was obtained for higher temperatures.

Figure 1.23 shows that the density curves of saturated solutions with $x = 0.8$ intersect at a temperature of about 35°C, which is close to the temperature point O in figure 1.20.

(b) *Conductivity*

The conductivity of solutions was determined by the author in experiments at a frequency of 1 kHz. Some experimental results for solutions of constant concentration at different temperatures are shown in figure 1.24. One can see that the specific conductivity value, χ, increases linearly with temperature. All the data available were fitted by the formula

$$\Lambda = R + rt - (S + st)\sqrt{M \times 10^3} \qquad (1.23)$$

where $\Lambda = \chi/M$ is the molar conductivity ($\Omega^{-1}\,\mathrm{cm}^2\,\mathrm{mol}^{-1}$), $M = c^*/m$ is the solution molarity ($\mathrm{mol\,cm}^{-3}$), and m is the salt molecular weight. From the density formula we have

$$c^* = \frac{a - a_1 t}{(100/c) - b + b_1 t}.$$

Substituting this relation into the expression for Λ (equation (1.23)) we then calculated the values of R, r, S and s by the least-squares method (see table 1.11). The dependences $\chi(t)$ calculated in this way are plotted by full curves in figure 1.24. In this figure one can see that the calculations agree quite satisfactorily with the experimental data. Differentiating equation (1.23) with respect to M and taking the derivative to be equal to zero, we can find the solution molarity, for which the specific conductivity at a given temperature has the maximum

$$10^3 M_{\max} = [2(R + r)/(S + s)]^2.$$

Now, using the formula for c^*, the corresponding values of c can be calculated. The calculation shows that for a solution in H_2O at 25°C the maximum value of χ corresponds to the 44.2 wt % concentration, and corresponds to 48.9 wt % at 50°C. For solutions in D_2O, these values are 45.6 wt % and 47.5 wt %, respectively. The specific and molar conductivities of saturated solutions are shown in figure 1.25. Both values change almost linearly with temperature, the conductivity of deuterated solutions being less in spite of

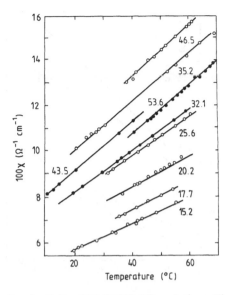

Figure 1.24 Conductivity of Rb(H,D)$_2$PO$_4$ solutions. The concentration is denoted at the curves in wt %. ○, $x = 0$; ●, $x = 0.8$.

Table 1.11 Constants of equation (1.23) for the molar conductivity of Rb(H,D)$_2$PO$_4$ solutions in water (M in mol l^{-1}).

	$x = 0$	$x = 0.8$
R	62.9 ± 2.6	42.8 ± 1.2
r	1.55 ± 0.08	1.46 ± 0.004
S	25.5 ± 1.7	15.3 ± 0.6
s	0.43 ± 0.05	0.46 ± 0.02
σ_0	0.97	0.23
Number of experiments	93	65
Temperature range (°C)	20.3–46.8	14.0–49.7
Concentration range (wt %)	15.2–49.4	32.1–54.1

the greater solubility of the salt. A comparison of the solubility values with those of solution concentrations corresponding to the maximum specific conductivity shows that the specific conductivity of saturated solutions decreases due to increasing concentration and increases with rising temperatures for the deuterated salt at temperatures above 0°C, and for the non-deuterated salt at temperatures above $\simeq 25$°C. The temperature contribution prevails (as can be seen from figure 1.25). Therefore in this case (cf. the case of the ammonium salt) it is in fact impossible to reach any certain conclusions about solution supersaturation (during the crystal growth process) on the basis of conductivity measurements.

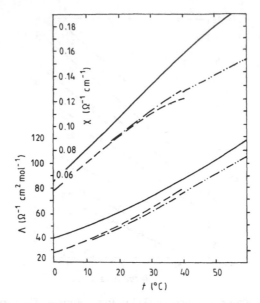

Figure 1.25 Conductivity and molar conductivity of $Rb(H,D)_2PO_4$ saturated solutions: ——, $x = 0$, for all the other curves $x = 0.8$; –––, tetragonal phase; –·–, phase II; –··–, phase I.

(c) *Freezing temperatures of solutions*

Solutions of known concentrations were placed in a cell with a semiconductor cooler and a stirrer. The system was cooled and heated at the rate of $0.5°C$ min^{-1} (the temperature being measured with a thermometer). As a rule, a degree of supercooling of the solution was observed, therefore the crystallization temperature was determined by extrapolating the kinetic curve (to an accuracy of $\pm 0.1°C$). The results are given in table 1.12.

Figure 1.26 shows (in the molal scale: moles per 1 kg of water) the decrease of the freezing temperature of water in RbH_2PO_4, KH_2PO_4 and $NH_4H_2PO_4$ solutions of various concentrations. One can see that up to a value of $\simeq 1.5$ moles per 1 kg H_2O, the decrease in the freezing temperature depends linearly on the concentration and does not depend on the salt. The ratio of the angular coefficient of the curve represented in figure 1.26 to the cryoscopic constants of H_2O (2.45/1.86) is equal to $\simeq 1.31$. Hence, the degree of dissociation of the salts is $\simeq 30\%$ and at these temperatures it does not actually depend on the solution concentration. It was not possible to construct a similar pattern for deuterated salts, since the minimum salt concentration was too high in the author's experiments and there appeared to be a mistake in the experiments made by Barkova and Lepeshkov (1968), since according to their data, a solution with 5.1 wt % of salt froze at $3.9°C$, i.e. at a temperature higher than the crystallization temperature of pure D_2O ($3.82°C$—see, for example, Brodskii (1952)).

Table 1.12 Freezing temperature of $Rb(H,D)_2PO_4$ solutions.

	$x = 0$			$x = 0.8$	
c (wt %)	t (°C)	Solid phase	c (wt %)	t (°C)	Solid phase
0	0.0	ice	0	3.0	ice
5.2	−0.7	ice	18.1	0.3	ice
10.6	−1.6	ice	20.2	0.0	ice
15.2	−2.3	ice	23.1	−0.4	ice
22.5	−3.4	ice	27.6	−1.0	ice
28.5	−4.5	ice	29.4	−1.2	ice
30.5	−5.1	ice + RbH_2PO_4	33.6	−1.9	ice + phase II
32.2	−1.0	RbH_2PO_4	34.9	0.8	phase II
12.36	−2.6	ice + KH_2PO_4[a]	20.0	0.9	ice + KD_2PO_4[a]
17.25	−4.2	ice + $NH_4H_2PO_4$[b]			

[a] Barkova and Lepeshkov (1968).
[b] Polosin (1946).

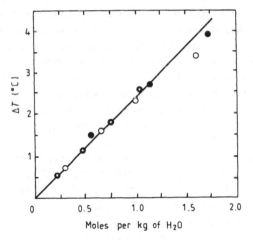

Figure 1.26 Decrease of the freezing temperature of water (ΔT) in solutions of different concentrations. ○, RbH_2PO_4; ◎, KH_2PO_4 (Barkova and Lepeshkov 1968); ●, $NH_4H_2PO_4$ (Polosin 1946).

1.4 The $Cs_2O-P_2O_5-(H,D)_2O$ system

For a long period, data on caesium dihydrogen phosphate were rather contradictory. Until the mid-1970s CsH_2PO_4 crystals were believed to be isomorphous to KH_2PO_4 crystals. Uesu and Kobayashi (1976) and Rashkovich *et al* (1977a,b) independently and practically at the same time showed that this idea was not correct. Data on the system and of the properties of $Cs(H,D)_2 PO_4$ solutions were quite limited, and those available were obtained mostly in experiments by the present author and co-workers.

1.4.1   Solid phase

$CsH_2 PO_4$ was first synthesized at the very beginning of the century (Berg 1901). Hendricks (1927) established the $CsH_2 PO_4$ crystals as being optically biaxial. In 1947 Matthias *et al* (1947) investigated the ferroelectric properties of the $KH_2 PO_4$ crystals. They believed that it was impossible to grow tetragonal $CsH_2 PO_4$ crystals because the great ionic radius of caesium would make such a structure unstable. Since those authors believed that $KH_2 PO_4$ crystals can possess ferroelectric properties only if they have tetragonal symmetry in the *para* phase, they thus believed that $CsH_2 PO_4$ could not be ferroelectric. However, in 1950 Seidl (1950) found that below 159 K $CsH_2 PO_4$ is ferroelectric and thus the idea proposed by Matthias *et al* (1947) proved to be mistaken. In 1948 Magyar (1948) carried out an x-ray examination of $CsH_2 PO_4$, determined the elementary cell parameters and related the crystals to orthorhombic symmetry. In 1952, Fellner-Feldegg (1952) produced x-ray patterns of rotation around the main crystallographic directions. Taking into consideration the extinction rule, the type of x-ray patterns and the possible atomic arrangements, he calculated the atomic coordinates for an elementary cell in a non-centrosymmetric space group C_2^{14} (mm2 point group).

In all the works mentioned above it was stated that tetragonal $CsH_2 PO_4$ crystals did not exist. However, later on this fact was forgotten and in some works (see, for example, a detailed monograph by Jona and Shirane (1962)) it was reported with reference to the same papers that $CsH_2 PO_4$ is isomorphous to $KH_2 PO_4$.

In 1975 Levstik *et al* (1975) repeated the experiments of Seidl on the temperature dependence of the permittivity and studied this dependence in $CsD_2 PO_4$ crystals. Following Fellner-Feldegg's example, they considered the crystals to be orthorhombic. The physical properties of $CsH_2 PO_4$ and $CsD_2 PO_4$ are treated in works by Rashkovich and Meteva (1978), Baranov *et al* (1979) and in numerous later papers.

A detailed study of crystal structure was published in 1976 by Uesu and Kobayashi (1976). They pointed out that Fellner-Feldegg had not noticed the difference in intensity of a number of diffraction reflections, and showed that the C_{2n}^2 (P2/m point group) is the correct space group. Thus, $CsH_2 PO_4$ crystallizes in monoclinic symmetry and possesses a symmetry centre at room temperature. Semmingsen *et al* (1977) and Nelmes and Choudhary (1978) reported the results of neutron diffraction analyses of $CsH_2 PO_4$ and $CsD_2 PO_4$.

The aim of our experiments was to study the possibility of growing $CsH_2 PO_4$ tetragonal crystals by varying the temperature and the content of CsOH and $H_3 PO_4$ in solution.

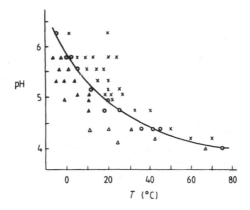

Figure 1.27   Effect of crystallization conditions on the habit of CsH₂PO₄ crystals. △, rhombic plates; ×, prisms; ○, full-faced forms.

Figure 1.27 shows the results of the experimental series in which spontaneous crystallization in small volumes of various acidity was observed. The value of the pH of solutions saturated at 25°C is plotted as the *y* axis. Solutions in contact with crystals were kept for several hours at the experimental temperature, then the liquid phase was decanted into a cell thermostatted to the same temperature. Crystallization proceeded under isothermic conditions due to evaporation of the solvent. It was found that in the experimental temperature range (−5–+80°C) and in the range of solution pH used (≃ 2–8) one monoclinic phase crystallizes, with crystals having a different habit depending on the experimental conditions.

In the region above the curve shown in figure 1.27, the crystals have the shape of prisms elongated along the *b* axis and bounded by {100}, {001} pinacoids and {110} vertical prisms (see figure 1.28(*a*)). In the region below the curve, the rhombic plates crystallize. This is analogous to the effects observed in the experiments of Berg (1901), Seidl (1950) and Bykova *et al* (1968). They are bounded by a well developed front {100} pinacoid and the faces of a {11$\bar{1}$} rhombic prism and a {011} longitudinal prism (figure 1.28(*b*)). The crystals grown under the conditions corresponding to the region near the curve of figure 1.27 have the faces of a {10$\bar{1}$} transverse pinacoid as well as all the faces mentioned above (see figure 1.28(*b*)).

In growing crystals by the solvent evaporation or lowering temperature techniques, the same regularities are observed. Thus, the crystals, which are several centimetres long and were grown at 35–40°C from a solution of pH ≃ 4.3 had the {100} and {001} front and base-pinacoid faces most developed, while the vertical prismatic faces and those of the transverse pinacoid were smaller than the other prismatic faces. The growth rate in the

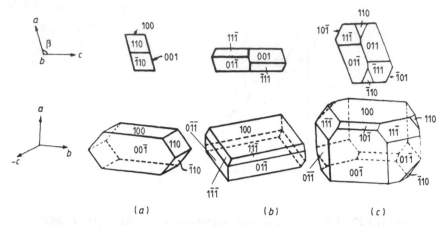

Figure 1.28 Orthogonal projection and form of CsH_2PO_4 crystals which crystallize under the conditions shown in figure 1.27: (*a*) above the curve; (*b*) below the curve; and (*c*) near the curve.

[010] direction was twice as great as that along [100] and three to four times greater than that along [001].

A similar situation was observed in the growth of crystals from deuterated solutions (the maximum deuteration degree being 98%).

Goniometric measurements of the full-faced crystals showed that the monoclinic angle and the axial ratios coincide with the data obtained by Uesu and Kobayashi (1976).

It should be noted that monoclinic symmetry only appeared in full-faced crystals. The crystals shown in figures 1.28(*a*) and (*b*) can also be described within the mm2 class of orthorhombic symmetry. The fact is that the accuracy of the goniometric measurement is not quite sufficient to allow an unambiguous choice between the mm2 and 2/m classes. Indeed, in an orthorhombic setting the angles (α_r), between the $(\bar{1}00)_r$ and $(10\bar{1})_r$ faces, and the $(100)_r$ and $(101)_r$ faces should be equal, with $\alpha_r = 180 + \tan^{-1} c_r/a_r = 108.14°$ (using Fellner-Feldegg's (1952) parameters). In the monoclinic setting ($a_r = c_m$, $b_r = b_m$), the first pair of faces will have indices $(100)_m$ and $(001)_m$, the angle between them being equal to $\beta = 107.742°$ (according to data by Uesu and Kobayashi (1976)). The second pair of faces will have indices $(\bar{1}00)_m$ and $(\bar{1}01)_m$ (see figure 1.28). The angle between these faces can be obtained from

$$\cos \alpha_m = \frac{1/a_m + \cos \beta / c_m}{(1/a_m^2 + 1/c_m^2 + 2 \cos \beta / a_m c_m)^{1/2}} \qquad \alpha_m = 108.137°.$$

Thus, α_m differs from β by 24', and $\alpha_m = \alpha_r$. This difference between α_m and β (due to the imperfection in the real crystal faces) can always be accounted for by the experimental error. A similar situation is observed for the faces of rhombic and longitudinal prisms, which can be taken to be the faces of the upper and lower rhombic pyramid.

The habit of the full-faced crystals unambiguously allows the establishment of the 2/m symmetry. The presence of a symmetry centre was confirmed by the absence of any piezoelectric effect along the three crystallographic axes at room temperature (Rashkovich and Meteva 1978). Pyroelectric effect measurements were carried out along the same directions at temperatures ranging from room temperature to that of liquid nitrogen and a pyroelectric effect was observed only along the *b* axis below the ferroelectric phase transition temperature.

Crystals are characterized by perfect cleavage along (100). Note that tetragonal crystals of monosubstituted phosphates have no such perfect cleavage.

1.4.2 Solubility polytherms

Zvorykin and Ratnikova (1963) determined the solubility of CsH_2PO_4 in water at 25°C as 66.26 wt %. Bykova *et al* (1968) studied the CsH_2PO_4 solubility in the temperature range 0–80°C. They prepared the salt by neutralizing caesium carbonate with orthophosphoric acid up to pH = 4.5. Data on the stoichiometry of the solution obtained from this salt are not given. The results of Bykova *et al* (1968) have been processed and the following expression for the solubility of CsH_2PO_4 in the range 25–80°C has been obtained

$$c = (53.7 \pm 0.6) + (0.224 \pm 0.009)t \pm 0.4. \qquad (1.24)$$

Hence, at 25°C $c = (59.3 \pm 0.4)$ wt % (the experimental value being 59.58 wt %). Our experiments with a solution of stoichiometric composition yielded $c = (58.7 \pm 0.3)$ wt %. In stoichiometric solutions of different deuteration degree the solubility of $Cs(H,D)_2PO_4$ in wt % at 25°C is practically independent (according to our data) of the deuteration degree.

In a closed container, CsH_2PO_4 melts incongruently at 300°C (Zhigarnovskii *et al* 1984). In air, according to Nirsha *et al* (1982) (see also Gupta *et al* 1980) three dehydration stages are observed: stage I (at 175–225°C) terminates by the formation of pyrophosphate ($Cs_2H_2P_2O_7$); stage II (at 135–285°C) terminates by the transition of pyrophosphates into polyphosphates; and at stage III (at 285–485°C) water is removed from polyphosphates, about 10% of the water being lost at even higher temperatures. The dehydrated salt being subsequently dissolved, a gelatinous solution is produced. $Cs(H,D)_2PO_4$ cannot crystallize from this solution at any value of pH. The irreversibility of dehydration (in contrast to KH_2PO_4) makes it impossible to use the dehydrated salt in the preparation of solutions with a high deuteration degree by dissolving it in D_2O.

1.4.3 Solubility isotherms

Solubility isotherms were studied by Rashkovich *et al* (1977a) at 25°C and 50°C in solutions with a deuteration degree of 0% and 98 ± 1%.

The experimental results are represented in tables 1.13 and 1.14 and in figures 1.29 and 1.30. In figure 1.29 one can see that the alkali and acid branches of the equilibrium curves in the $Cs_2O-P_2O_5-H_2O$ system are nearly linear, the alkali branch being roughly parallel to the y-coordinate. Thus, the solubility of CsH_2PO_4 (in wt %) does not depend on the alkali concentration in the solvent. To be precise, this branch is slightly bent towards

Table 1.13 Solubility isotherms of CsH_2PO_4 in the $Cs_2O-P_2O_5-H_2O$ system (wt %)[b].

pH	Cs₂O	P₂O₅	Cs₂O	P₂O₅
	25°C		39°C	
7.7	51.6	17.8	42.8	18.3
7.2	48.2	17.6	38.2	19.4
6.65	45.0	17.5	32.4[a]	30.6[a]
6.4	43.4	17.4	44.7°C	
6.05	41.8	17.4		
5.6	39.5	17.5	43.5	18.7
5.1	37.9	17.8	39.0	19.9
5.0	37.3	17.6	33.4[a]	31.7[a]
5.0	37.2	17.9	50°C	
4.1	36.1	18.5	61.1	19.4
3.9	36.0	18.9	53.3	19.1
3.6	36.4	20.0	50.6	19.1
2.85	38.5	24.5	47.9	19.2
2.7	38.2	24.8	44.0	19.0
2.6	38.9	26.2	43.0	19.6
2.2	39.9	28.8	42.3	19.8
2.1	38.9[a]	28.8[a]	40.7	19.4
1.7	34.8[a]	28.4[a]	40.2	20.5
1.3	29.9[a]	27.9[a]	40.4	21.1
	33.2°C		40.6	21.3
			40.8	21.9
	42.2	17.8	41.5	26.2
	37.2	19.0	41.3	29.2
	31.6[a]	29.5[a]	39.9[a]	32.3[a]
			38.6[a]	32.4[a]

[a] The ultra-acid salt is in equilibrium with the solution.

[b] Linear approximation constants of the isotherms are given in table 1.16.

Table 1.14 Solubility isotherms of CsD$_2$PO$_4$ in the Cs$_2$O–P$_2$O$_5$–D$_2$O system (wt %)[b].

25°C		50°C	
Cs$_2$O	P$_2$O$_5$	Cs$_2$O	P$_2$O$_5$
46.2	17.6	42.2	19.1
39.7	17.6	40.0	19.6
38.3	17.9	39.5	21.8
36.8	20.8	39.6	21.9
37.6	22.7	39.5	22.6
37.7	23.5	40.3	24.5
37.9	23.7	40.5	25.0
38.0	24.2	40.4	25.1
37.9	24.3	40.8	26.4
38.6	25.8	38.0[a]	31.7[a]
36.8[a]	28.3[a]		

[a] The ultra-acid salt is in equilibrium with the solution.
[b] Constants of the linear equations of the isotherms are given in table 1.16.

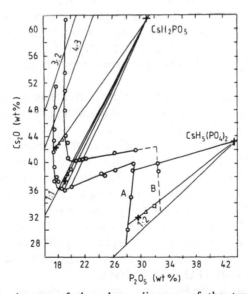

Figure 1.29 A part of the phase diagram of the ternary system Cs$_2$O–P$_2$O$_5$–H$_2$O. Temperature: A, 25°C; B, 50°C; +, 33.2°C; △, 39°C; □, 44.7°C. The circles plotted on the lines passing through the solid phase compositions correspond to experiments with the same solvent composition.

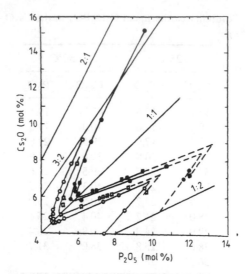

Figure 1.30 Phase diagram of parts of the $Cs_2O-P_2O_5-H_2O$ system
($\bigcirc$), and of the $Cs_2O-P_2O_5-D_2O$ system ($\square$). Open symbols refer to
data at 25°C, full symbols refer to data at 50°C.

the y axis, i.e. the solubility of CsH_2PO_4 in water containing alkali
first increases slightly as the alkali concentration increases, reaching a
minimum in water with $\simeq 19$ wt % Cs_2O, and then increases slightly. The
position of the curve shows that even if compounds with molar
ratios of $Cs_2O/P_2O_5 = 4:3$ and $3:2$ ($Cs_2HPO_4.2CsH_2PO_4.nH_2O$ and
$Cs_2HPO_4.CsH_2PO_4.nH_2O$) do exist, they dissolve in pure water
incongruently.

The acid branch of the equilibrium curve has a noticeable positive
inclination and, accordingly, an increasing solubility of CsH_2PO_4 is observed
with increasing solution acidity.

At the invariant point corresponding to the coexistence of the solution
and solid CsH_2PO_4 and $CsH_5(PO_4)_2$ at 25°C, the solution contains
39.9 wt % Cs_2O and 28.9 wt % P_2O_5. The ultra-acid salts dissolve congruently
in water, the saturated solution containing 64.2 wt % $CsH_5(PO_4)_2$.

At 50°C the $Cs_2O-P_2O_5-H_2O$ system behaves in much the same way.
The solubility of CsH_2PO_4 in water is 66 wt %. By extrapolating the data
obtained it was found that in a three-phase equilibrium with the ultra-acid
salt the concentration of Cs_2O is 42.8 wt %, that of P_2O_5 is 32.2 wt %, and
the solubility of $CsH_5(PO_4)_2$ is 75.8 wt %.

In the wt % scale the inclination of the equilibrium curve acid branch
considerably decreases with increasing temperature (see figure 1.29), and
accordingly, the effect of the solution acidity on solubility also decreases.

The isotherms under study are characterized by the solubility minimum
of CsH_2PO_4 being appreciably shifted from the stoichiometric solution into

the alkali region, this shift becoming greater with increasing temperature. It is worth noting that the compositions of solutions that are in equilibrium with $CsH_5(PO_4)_2$ lie on the straight line passing through the composition point of CsH_2PO_4.

Solubility isotherms of the $Cs_2O-P_2O_5-D_2O$ system are shown in figure 1.30 in mol %. For comparison, the data on the system using H_2O are also given in this figure.

Figure 1.30 shows that the behaviour of both systems is analogous. The equilibrium curves for the system with D_2O are above those for the system with H_2O, the difference in solubility of CsH_2PO_4 and CsD_2PO_4 becoming less with increasing temperature. All the curves are roughly parallel.

It should be noted that, in the mol % scale, the solubility of mono-substituted caesium orthophosphates increases significantly with increasing alkali concentration above a certain limiting (although not high) value. It should be recalled that in the wt % scale the solubility of CsD_2PO_4 in D_2O at 25°C is actually the same as that of CsH_2PO_4 in H_2O at that temperature.

At 50°C the solubility of CsD_2PO_4 in D_2O in wt % is less than that of CsH_2PO_4 in H_2O and is equal to $\simeq 64$ wt %. As we have seen above, in this concentration scale the solubilities in D_2O of mono-substituted orthophosphates of other alkali metals and ammonium are much higher than those in H_2O.

To test the hypothesis of the composition inconstancy of mono-substituted caesium orthophosphate due to a solubility minimum shift, crystals grown under different conditions were subjected to thorough chemical analysis. The crystals were grown by the lowering temperature technique from solutions with the molar ratio Cs_2O/P_2O_5 being equal to 1.25, 0.95 and 0.88. The density of these crystals, determined by the hydrostatic weighing of single crystal samples of about 4 g in toluene, was found to be 3.230 ± 0.001 g cm^{-3} for all samples. Chemical analysis was carried out by means of potentiometric titration of solutions prepared from the crystals. For all three crystals the molar ratio Cs_2O/P_2O_5 was 1 ± 0.005, the calculated accuracy of this ratio determination being ± 0.01. It was found, however, that the pH of solutions prepared by dissolving the crystals in a 50-fold amount of water decreased regularly and was 4.59, 4.51 and 4.47 (± 0.05), respectively. Assuming this pH variation to be connected with changes in composition, it is possible to estimate these composition changes. Differentiating the formula given below for the titration curve we find at the first point of equivalence

$$\Delta y = \Delta(Cs_2O/P_2O_5) = \frac{\Delta(pa_H)}{1 + 0.5(K_1\gamma_0\gamma_2/K_2\gamma_1)^{1/2}}$$

(the notation being described in section 1.4.4(b)). Assuming that in diluted solutions the ratio of the activity coefficients is close to unity and taking into consideration that at 25°C $K_1/K_2 = 11.28 \times 10^4$, we obtain $\Delta y \simeq$

$10^{-2}\,\Delta(pa_H)$. In our experiments $\Delta(pa_H) \simeq \pm 0.05$ and therefore Δy could be $\simeq 0.0005$. Unfortunately, the accuracy of the analytical methods available is much less than this limit.

1.4.4 Properties of solutions

Data on solutions of $Cs(H,D)_2PO_4$ are quite limited, so only the results from the author's work are presented.

(a) Density

The density of saturated solutions of CsH_2PO_4 in H_2O was determined by the hydrostatic weighing of a quartz plate. The suspension was stirred with a magnetic agitator. At a given temperature, equilibrium was reached in 1–1.5 hours and was established by constancy of the plate weight for 10 minutes. Weight measurements were made with the agitator being turned off.

The results of the density measurements in the temperature range 14–54°C are given in figure 1.31. Processing the data by the method of least squares yields

$$d = (1.637 \pm 0.002) + (0.00504 \pm 0.00006)t \pm 0.002. \tag{1.25}$$

(b) The effect of the solvent composition

The effect of the solvent composition on the pH of the saturated solution and on the solubility of CsH_2PO_4 will be discussed here in more detail because these parameters are important for practical crystal growing. Relevant experimental data are given in table 1.13.

When a tribasic acid is neutralized by an alkali, the hydrogen ion activity can be defined from the following equation (Robinson and Stokes 1959)

$$(3-y)K_1K_2K_3/\gamma_3 + (2-y)K_1K_2a_H/\gamma_2 + (1-y)K_1a_H^2/\gamma_1 = ya_H^3/\gamma_0$$

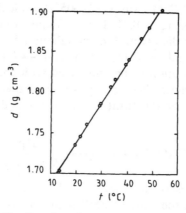

Figure 1.31 Density of saturated solutions of CsH_2PO_4.

where:

K_1, K_2, K_3 are the acid dissociation constants

$\gamma_0, \gamma_1, \gamma_2, \gamma_3$ are the molar coefficients of acid residual activity, the residual negative charge value being denoted by subscripts

y is the ratio of alkali gram-equivalents to acid gram-equivalents

a_H is the hydrogen ion activity measured by a pH meter ($pa_H = -\log a_H$), previously and less strictly the value of pa_H was called pH.

The dissociation contants of orthophosphoric acid differ by four to five orders (at 25°C $pK_1 = 2.148$ (Bates 1951); $pK_2 = 7.201$ (Gary *et al* 1964); and $pK_3 \simeq 12$), therefore equation (1.25) can be simplified and we can write approximate relations that are valid even at some distance from the points of equivalence. Thus, for $y < 1$

$$pa_H = \log[y/(1-y)] + pK_1 + \log(\gamma_1/\gamma_0)$$

for $1 < y < 2$

$$pa_H = \log[(y-1)/(2-y)] + pK_2 + \log(\gamma_2/\gamma_1)$$

and at the first point of equivalence (where $y = 1$) we have

$$pa_H = 0.5[pK_1 + pK_2 + \ln(\gamma_2/\gamma_0)] = 4.675 + 0.5\ln(\gamma_2/\gamma_0).$$

The experimental data (table 1.13) can be described by these relations, and in this way we can calculate the ratios of the corresponding activity coefficients, which differ from unity in concentrated solutions and depend on the solution composition (i.e. on the value of y).

Figure 1.32 represents the dependence of the pH of solutions saturated at 25°C on the molar ratio Cs_2O/P_2O_5 in solution (note that by pH we mean, as before, pa_H). One can see that experimental values of the pH depend linearly on the logarithm of Cs_2O/P_2O_5. For a solution of stoichiometric composition we obtain pH $\simeq 4.35$ which yields $\gamma_2/\gamma_0 = 0.23$.

Data on the ratios of the phosphoric acid ion activity coefficients for other values of y are not provided here, because no other data are available for comparison.

Figure 1.32 also demonstrates the dependence between the content of free alkali or acid and the pH of saturated solutions. The most abrupt changes in pH are observed, as one would expect, with low concentrations of acid or alkali in water.

Note that changes in the pH with rising temperature (for saturated solutions) are appreciable. The effect is enhanced with decreasing solution acidity. Thus, for instance, in the temperature range 5–50°C a temperature increase of 10°C increases the pH by $\simeq 0.2$ for a solution in water, and by $\simeq 0.3$ for a solution in water containing 12.5 wt % Cs_2O.

To conclude, a plot of the $Cs(H,D)_2PO_4$ solubility as a function of the content of acid or alkali in the solvent at 25°C and 50°C, which can be used

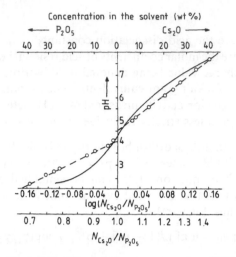

Figure 1.32 pH of CsH_2PO_4 solutions saturated at 25°C: – – –, effect of the solution composition; ——, effect of the solvent composition.

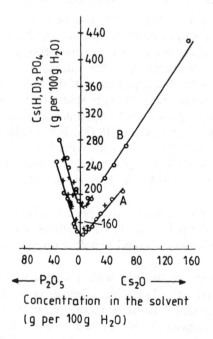

Figure 1.33 Effect of the solvent composition on $Cs(H,D)_2PO_4$ solubility at 25°C (A) and at 50°C (B): ○, solutions in H_2O; +, solutions in D_2O.

in practice, is provided (figure 1.33). All concentrations are given in g per 100 g H_2O and one can clearly see all the peculiarities mentioned previously from the solubility isotherms in the system being discussed; namely, the shift of the solubility minimum into the alkali region, the linear increase of the salt solubility with increasing concentration of an alkali or acid in the solvent and the closeness of the solubilities of mono-substituted caesium orthophosphate in solutions of ordinary and heavy water in this concentration scale (at 25°C the solubility in D_2O is several times greater than that in H_2O and at 50°C the reverse is the case).

1.5 The $Cs_2O-As_2O_5-(H,D)_2O$ system

The literature data on this system are quite scant. $Cs(H,D)_2AsO_4$ crystallizes in tetragonal phase, its habit being similar to that of the tetragonal crystals of the KH_2PO_4 group.

The solubility of CsH_2AsO_4 was studied by Shklovskaya and Arkhipov (1967). No data on the solution stoichiometry were given in their work. Here are their results:

t (°C)	0	25	50	75
c (wt %)	66.62	71.63	76.23	79.92

Balascio *et al* (1975) reported the following data on the solubility of CsD_2AsO_4:

t (°C)	20	30	40	50	60
c (wt %)	77.53	79.17	80.73	82.82	84.42

These results are quite well approximated by our relation

$$c = (73.96 \pm 0.35) + (0.174 \pm 0.008)t \pm 0.15. \qquad (1.26)$$

Loiacono *et al* (1977) described the process of growing CsD_2AsO_4 crystals. Because of the very high solubility of the salt, they used a mixture of D_2O with 70 wt % of deuterated ethanol as a solvent, which allowed them to reduce the solubility and to obtain crystals of high quality. In the work of Loiacono *et al* (1976) it is noted that it is difficult to grow crystals with a deuteration degree of more than $\simeq 85\%$ due to the formation of a monoclinic phase. To be more exact, the authors only make a suggestion concerning the crystallization of that phase giving no evidence to confirm this suggestion.

The dehydration of caesium orthoarsenate was studied by Gallagher (1976), Zhigarnovskii *et al* (1984) and Rakhimov *et al* (1985). In a sealed container the salt melts congruently at 284°C ($x = 0$) and at 295°C ($x = 0.95$). In air the weight loss begins at $\simeq 170$°C, the decomposition endothermic peak on the curves of the differential thermal analysis is at 260–270°C.

The author studied the solubility isotherm at 55°C in the $Cs_2O-As_2O_5-H_2O$ system (Rashkovich 1979) and used the caesium dihydroarsenate salt of special purity using a molar ratio $Cs_2O/As_2O_5 = 0.87$. Surplus acid appeared to be in the reagent in the form of a considerable amount of ultra-acid salt ($\simeq 21$ wt %). The solution of this 'caesium dihydroarsenate' saturated at 55°C had a pH $= 3.45$. Initial reagents also used were H_3AsO_4 with a concentration of 72 wt % and CsOH with a concentration of 53.5 wt %. Solutions of different acidity with undissolved solid phase were placed into test tubes of about 40 cm^3, which in turn were placed into a thermostat with a shaking device. The solutions were overheated until the solid phase was completely dissolved; then the temperature was lowered to 55°C and was kept constant to an accuracy of ± 0.1°C. The solutions were stirred for 24 hours (6–8 hours are required to establish equilibrium) and were then allowed to settle. Then the solution samples were diluted with water 1:(20–50) and titrated with NaOH at room temperature by passing nitrogen bubbles through the solution. Alkaline solutions were initially acidified with a certain amount of H_3AsO_4. The points of equivalence (with pH $\simeq 4.5$ and 8.8) were determined to an accuracy of ± 0.05 ml, and as a result the error in the determination of the concentrations of Cs_2O and As_2O_5 was about 1%. Crystals were identified under a microscope, the composition of the solid phase being determined from the results of a potentiometric titration of a solution prepared by dissolving the crystals in water.

The experimental results are given in table 1.15 and in figure 1.34. Figure 1.34 shows that the equilibrium curve differs greatly in appearance from the curves discussed in the previous sections; no kinks in the solubility of CsH_2AsO_4 in pure water are observed and the curve shape is far from linear. The solubility of CsH_2AsO_4 in H_2O, determined by interpolating our

Table 1.15   Isotherm at 55°C in the $Cs_2O-As_2O_5-H_2O$ system (wt %).

Liquid phase		Solid phase	Liquid phase		Solid phase
Cs_2O	As_2O_5	Cs_2O/As_2O_5 (mol)	Cs_2O	As_2O_5	Cs_2O/As_2O_5 (mol)
20.8	42.6	1:2	40.4	31.1	1:1
23.2	41.6	1:2	41.4	30.2	1:1
31.0	39.5	1:2	41.7	29.7	1:1
38.1	39.5	1:2[a]	44.2	28.5	1:1
36.1	38.8	1:1	49.1	27.4	1:1
37.3	36.2	1:1	57.3	22.0	2:1
38.8	32.4	1:1	39.7	21.2	1:1[b]
39.4	31.9	1:1			

[a] Metastable equilibrium.

[b] Equilibrium was established at 25°C.

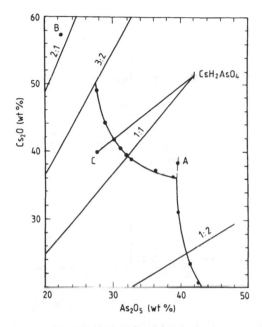

Figure 1.34 Phase diagram of a part of the Cs_2O–As_2O_5–H_2O system at 55°C. A, metastable equilibrium with the ultra-acid salt; B, dibasic salt in equilibrium; and C, equilibrium at 25°C.

data, is 76.3 ± 0.2 wt % at 55°C, which differs slightly from the value of 77.2 wt % from the interpolated data of Shklovskaya and Arkhipov (1967). At the invariant point, corresponding to the coexistence in solution of dihydrogen arsenate and ultra-acid salt, the solution contains about 36.0 wt % Cs_2O and 39.5 wt % As_2O_5. The ultra-acid salt dissolves congruently in water, its solubility at 55°C being $\simeq 74.1$ wt %.

Table 1.15 and figure 1.34 also give the result of one experiment at 25°C (the solution containing 41.7 wt % Cs_2O and 29.7 wt % As_2O_5 at 55°C and having a pH of 6.2 being cooled to this temperature).

Figure 1.35 shows the solubility of CsH_2AsO_4 as a function of the content of acid or alkali in the solvent (cf. figure 1.33). One can clearly see that the choice of concentration scale is essential in describing the qualitative behaviour of the salt solubility.

1.6 Liquidus surface of phase diagrams

In this section we will discuss the general features of the phase diagrams studied as well as the differences between them due to the types of cations and solvents.

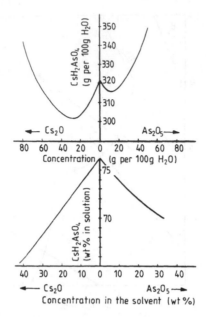

Figure 1.35 The effect of the solvent composition on the solubility of CsH_2AsO_4 at 55°C.

It has already been noted that most parts of both branches of the solubility isotherms of orthophosphates are linear

$$c_1 = a + bc_2 \tag{1.27}$$

where c_1 is the alkaline oxide concentration and c_2 is the concentration of P_2O_5. The constants for equation (1.27) are given in table 1.16 for all isotherms discussed.

One can see from table 1.16 that the alkali branches of the isotherms are not approximated by linear relations as closely as the acid ones are. In the case of the caesium salt, the mean square errors are especially large.

Before proceeding to the comparison of the data obtained, the choice of a reasonable concentration scale should be discussed. Its importance becomes evident when, for example, the solubility of CsH_2PO_4 in H_2O (L_H) and that of CsD_2PO_4 in D_2O (L_D) are compared. It has been mentioned above that in the wt % scale in both cases the solubilities at 25°C are equal and amount to 58.7 wt %. In table 1.17 these data are represented in other concentration scales.

The data in table 1.17 show that the solubility isotope effect can be taken to be either zero, positive or negative depending on the scale used. It should also be noted that the linear dependence of the solubility on temperature, which holds true for all compounds under investigation in the wt % scale, turns out to be a linear-fractional one in other concentration scales. This is

Table 1.16 Constants of equation (1.27) for the solubility isotherms for mono-substituted ortho-phosphates of alkali metals and ammonium (concentrations in wt %).

Cation	x (%)	Acid branch			Alkali branch		
		a	b	σ_0	a	b	σ_0
(a) 25°C							
NH_4	0	6.02 ± 0.04	0.023 ± 0.001	0.04	−6.5 ± 0.2	0.74 ± 0.01	0.2
K	0	5.26 ± 0.08	0.154 ± 0.003	0.10	−8.1 ± 0.4	1.45 ± 0.02	0.5
K	98	6.8 ± 0.4	0.15 ± 0.01	0.30	−13.9 ± 0.8	1.66 ± 0.03	0.5
Rb	0	15.5 ± 0.4	0.30 ± 0.1	0.10	−40 ± 5	3.8 ± 0.3	0.6
Rb	90	22.3 ± 0.2	0.198 ± 0.008	0.09	−86 ± 10	5.8 ± 0.5	0.9
Cs	0	28.8 ± 0.6	0.39 ± 0.02	0.20	−390 ± 140	25 ± 8	2.6
Cs	98	29.7 ± 0.6	0.34 ± 0.02	0.90	315 ± 330	−16 ± 11	4.6
(b) 50°C							
NH_4	0	9.5 ± 0.2	0.019 ± 0.003	0.05	−19 ± 2	1.22 ± 0.08	0.7
K	0	9.35 ± 0.08	0.118 ± 0.003	0.10	−14.0 ± 0.4	1.58 ± 0.01	0.3
Rb	0	24.6 ± 0.1	0.172 ± 0.004	0.06	−129 ± 13	7.6 ± 0.6	0.4
Rb	90	26.3 ± 0.4	0.14 ± 0.01	0.10	−200 ± 30	10.4 ± 1.3	1.0
Cs	0	36.6 ± 0.4	0.19 ± 0.02	0.10	−530 ± 360	30 ± 19	4.6
Cs	98	32.9 ± 0.8	0.30 ± 0.03	0.10	—	—	—

Table 1.17 Solubility isotope effect of mono-substituted caesium ortho-phosphate at 25°C.

Concentration scale	Solubility		$L_D - L_H/L_H$ (%)
	L_H	L_D	
wt %	58.7	58.7	0
mol %	10.0	10.9	9.0
Moles of salt per mole of water	0.111	0.123	10.0
Moles of salt per 1 kg of water	6.18	6.13	−0.86

not the case as far as the solubility isotherms are concerned. Here the linear relationship between the concentrations for the equilibrium curve is conserved in the case when the formulae of transition from one concentration scale to another are linear-fractional functions with the same denominator for the concentration of each component. Several transition formulae are given in table 1.18.

Table 1.18 shows that the correlation between the values of the coefficients for different substances changes, although the linear connection between concentrations is conserved. It is not only the tangent value of the slope angle that can change, but the sign of this value can change also. For example, the solubility of $NH_4H_2PO_4$ in wt % at 50°C decreases with increasing content of acid in the solution, since $b < 0$ (see table 1.16). The situation is the same when the concentration is expressed in moles per kg of H_2O ($B < 0$), but in the mol % scale the solubility appears to increase ($T > 0$). For caesium systems the acid branch at 50°C behaves in the opposite way: $b > 0$ but $T < 0$.

Thus, the only invariant features conserved during the calibrating transformations are: first, the linearity of the equilibrium curve branches (deviations from linearity being conserved as well); second, the presence (or absence) of a singularity of the intersection point of both isotherm branches; and, third, the shift (or absence of shift) of the solubility isotherm minimum.

From the above considerations and from figure 1.36 (where some of the isotherms studied are given) it follows that the diagrams being discussed are characterized by: the linearity of the solubility branches of mono-substituted orthophosphates; by rather sharp solubility minima; and, in some cases, by the presence of the minimum shift from the line referring to the stoichiometric solution. What is the significance of these data? What properties of solutions and of solid phases do they reflect?

The contemporary views concerning the solubility isotherm shape are as follows.

At the general meeting of the chemical section of the Russian Academy of Sciences in April 1921 and at the 2nd Mendeleev Congress (Petrograd,

Table 1.18 Recalculation formulae for the coefficients of the solubility isotherm linear equations from the wt % scale into other concentration scales where m_1, m_2, m_3 are the molecular weights of Me_2O, P_2O_5 and H_2O, respectively.

	mol %	moles per 1 kg H_2O
	$N_1 = k + TN_2$	$M_1 = A + BM_2$
Free term	$k = \dfrac{100 m_3 a}{100 m_1 - c(m_1 - m_3)}$	$A = \dfrac{10^3}{m_1}\,\dfrac{a}{100 - a}$
	$\Delta k = \dfrac{k^2}{a^2}\,\dfrac{m_1}{m_3}\,\Delta a$	$\Delta A = \dfrac{A}{a}\,\dfrac{100}{100 - a}\,\Delta a$
Angular coefficient	$T = \dfrac{100 m_2 b - a(m_2 - m_1)}{100 m_1 - a(m_1 - m_3)}$	$B = \dfrac{m_2}{m_1}\,\dfrac{100 b - a}{100 - a}$
	$\Delta T = \dfrac{100 m_2 \Delta b + [m_2 - m_1 + T'(m_1 - m_3)]\Delta a}{100 m_1 - a(m_1 - m_3)}$	$\Delta B = \dfrac{100 m_2 \Delta b + (m_2 + Bm_1)\Delta a}{m_1(100 - a)}$

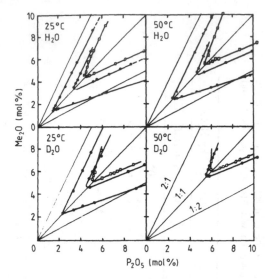

Figure 1.36   The solubility isotherms of mono-substituted orthophosphates: ○, potassium; ●, rubidium; □, caesium.

May 1922) N S Kurnakov gave a lecture devoted to the centenary of Claude Louis Berthollet. In this lecture he generalized his concepts concerning singularity points of chemical diagrams (Kurnakov 1925). Discussing the surface topology of phase diagrams he showed that a chemical individual should have a corresponding singularity point (line) on a liquidus or solidus surface (or on both), whose composition corresponds to that of the individual. If the substance is dissociated in the liquid state, the singularity point on the liquidus curve converts to a smooth maximum, whose position refers to that of the substance. If the individual does not produce solid solutions, the solidus line (surface) is contracted into a straight line parallel to the temperature axis. Dissociation in liquid and solid states gives rise to maxima on liquidus and solidus surfaces, their tangent points not necessarily corresponding to the stoichiometric composition of the solid phase. The shift of the maximum on the liquidus curve from the stoichiometric composition unambiguously indicates the formation of solid solutions. Kurnakov pointed out the essential difference between substances whose melting curve maxima correspond to their composition and those which are shifted from it. He called the former 'daltonides' (Kurnakov and Zhemchuzhni 1912) and the latter 'berthollides' (Kurnakov and Glazunov 1912).

Kurnakov's theory of singular phase diagram points was confirmed by numerous experiments by his co-workers. The results of those experiments were generalized in Kurnakov's monograph *Introduction in Physicochemical Analysis* (4th edition, 1940) and in the monograph by Anosov and Pogodin (1947).

By dissociation, Kurnakov and his co-workers meant not electrolytic dissociation, but the decomposition of a compound into its initial components. From a modern point of view, it corresponds to the different degree of phase ordering (or disordering), i.e. to the number of defects present. A small difference in energy producing defects (for example, in the cation or anion) shifts the thermodynamic potential minimum (corresponding to the fusibility curve maximum) towards those defect phases which are produced with a lower energy expenditure (Kröger 1964).

According to Kurnakov (*Introduction in Physicochemical Analysis* 1940, p 60), in ternary systems the liquidus surface corresponding to the crystallization of a given substance is a syncline or anticline surface (figure 1.37). The less the dissociation of a compound, the more distinct is the intersection angle of slopes, i.e. the crest line. The section of the liquidus surface produced by a plane perpendicular to the temperature axis yields the equilibrium curve, i.e. the solubility isotherm. As figure 1.37 shows, in the anticline case there is a minimum on the isotherm, which corresponds to the compound solubility in a pure solvent; in the syncline case there is a maximum. The isotherm extreme points form the liquidus surface crest lines. The crest lines (Mm) shown in figure 1.37 lie in the plane passing through the temperature axis and the composition point of a crystallizing compound. This plane is called a singular secant plane. The anticline surface crest line may be either singular or non-singular depending on whether a compound in a liquid state (in solution) has dissociated (figure 1.37(b)).

The section of the liquidus surface produced by a plane parallel to the temperature axis yields the solubility polytherm. In the anticline case the solubility minimum on the isotherm refers to the maximum on the corresponding polytherm. If a polytherm lies in the plane of a two-component system AB, the liquidus curve is obtained; various types of this curve have

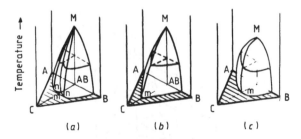

Figure 1.37 Syncline (*a*) and anticline ((*b*), (*c*)) (Kurnakov 1940). The crystallization surfaces of an AB compound in an A–B–C ternary system. The lines Mm are plotted in the singular secant plane. The relation of the A and B components corresponding to the minimum solubility points AB on the syncline isothermic sections (the crossing points of the isotherms with the Mn lines) depends on the temperature.

been discussed above. It is essential to point out that the characteristic points of isotherms correspond to those of polytherms.

Syncline surfaces are typical of some neutralization systems, e.g. when a salt is produced as a result of the interaction between a strong acid (hydrochloric, sulphuric or nitric acids) and a strong base (NaOH, KOH). The decreasing solubility of NaCl, KCl, NaNO$_3$ and others in solutions containing excess acid or alkali is due to the salting-out of the same ions and this follows from the mass action law.

In the cases under consideration, the anticline surface is formed. Figure 1.38 shows a part of the RbH$_2$PO$_4$ crystallization surface constructed upon Gibbs' triangle (in the mol % scale) and bounded by the isohydric line vertical plane at 84 mol % H$_2$O and the isothermic sections at 0°C and 75°C. The liquidus surface crest line and the 25°C and 50°C isotherms are plotted according to the data presented in this chapter; the 0°C and 75°C isotherms are plotted schematically. It is characteristic that, as the temperature increases, the crest line singularity disappears; it leaves the singular secant plane and is shifted into the alkali region. The sharp minimum of the solubility isotherm, as seen previously (figure 1.22), becomes smooth. A similar graph can be plotted for the Cs$_2$O–P$_2$O$_5$–(H,D)$_2$O system; in this graph the crest line shift would be even more pronounced. In systems with potassium and ammonium the crest line seems to lie in the singular secant plane at all temperatures (at least below 50°C).

Using extensive experimental material, Ravich (1940) studied in detail the reasons why the crest line can apparently lose its singularity, and subsequently introduced the notion of a mixed type of surface (see figure 1.39). He considered three main factors: (i) the distance of the isotherm being studied from the compound melting temperature; (ii) the electrolyte strength and its salting-out effect; and (iii) the formation of stable complexes in solution increasing the salt solubility.

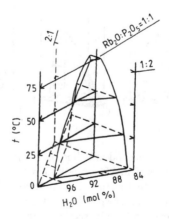

Figure 1.38 A part of the RbH$_2$PO$_4$ crystallization surface.

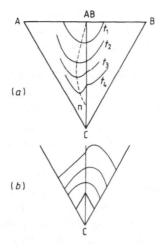

Figure 1.39 (*a*) The formation of a mixed surface (Ravich 1940). Elements characteristic of a syncline appear as temperature decreases from t_1 to t_4, but due to the weak salting effect of component B there is no line (AB)n in triangle (AB)BC; and (*b*) the types of isotherms in the solvent field. The linearity of isotherms in the ice field is characteristic of many systems.

If solid solutions are not produced, the maximum on the two-component system liquidus curve must correspond to the stoichiometric composition. Therefore as the temperature approaches the melting point, the syncline should transform into the anticline. If the temperature is approaching the melting point in the anticline case, its crest line should approach the singular secant plane. On the other hand, as we shall see below, with the temperature decreasing and approaching the ice field, the crest line also often approaches the singular secant plane.

If the salt is produced as a result of the interaction of two strong electrolytes, the syncline appears. If one electrolyte is much stronger than the other, the salting-out effect occurs only in that side of the diagram, where this electrolyte is redundant. It results in the production of a mixed surface and the solubility minimum does not correspond to the stoichiometric solution composition but is shifted towards the strong electrolyte. The stronger the electrolyte, the greater is the shift of the solubility minimum. The production of complexes in solution increases the salt solubility thereby inhibiting the salting-out process and the appearance of an implicit singularity.

In the solvent crystallization field the increasing electrolyte strength produces the opposite effect as compared to that in the salt crystallization field. In the latter case the maximum on the solubility isohydric-polytherm is shifted from the singular secant, and in the ice field the increasing electrolyte strength leads to an even greater decrease of the solvent freezing temperature which results in a much more pronounced crest-line singularity. The opposite effect is produced by the tendency to form complexes that in the ice field results in increasing the freezing temperature due to the decreasing total number of dissolved particles.

The above considerations present a fairly complete phenomenological view of the phase diagram liquidus surface in various systems. As for the phosphate systems the following data can be added. Ivakin and Voronova (1973) studied equilibrium in orthophosphoric acid aqueous solutions and proved the existence of a considerable number of dimers in concentrated solutions: $H_5(PO_4)_2^-$ $(H_2PO_4^-.H_3PO_4)$ in an acid medium, $H_4(PO_4)_2^{2-}$ $(H_2PO_4^-.H_2PO_4^-)$ in a neutral medium, and $H_3(PO_4)_2^{3-}$ $(H_2PO_4^-.HPO_4^{2-})$ in an alkali medium. Selvaratnam and Spiro (1965), who studied the mobility of the orthophosphoric acid ions, also confirmed the formation of $H_2PO_4^-.H_3PO_4$. Cerreta and Berglund (1987) showed that $NH_4H_2PO_4$ and KH_2PO_4 saturated solutions contained only 20% and 40% $H_2PO_4^-$ monomers, respectively, other anions being associated. Following van Wazer (1958), Lilich et al (1971) believe that $H_2PO_4^-$ ions are associated not in dimers but produce chain and ring structures, a considerable amount of water in the bound state being included into such a macrostructure of phosphoric acid. According to Lilich et al (1971), the formation of dimers under these conditions (particularly, $H_5(PO_4)_2$) should result in destroying the $H_2PO_4^-$ ion macrostructure and releasing water which can then dissolve an additional amount of salt. This effect is used by them to account for the increasing solubility of mono-substituted orthophosphates of alkali metals on the addition of H_3PO_4.

A thorough discussion of these ideas was presented in the work by Lilich and Chernykh (1970) which was based on Samoilov's (1957) considerations concerning the structure of electrolytes in aqueous solutions and the hydration of ions. Lilich and Chernykh (1970) considered the main types of behaviour of the salt solubility isotherms as a function of the concentration of the other electrolytes with the same ions in solution and suggested the following explanations for the four main types.

(i) Type I. The salt solubility decreases monotonically with increasing concentration of the second electrolyte. This is typical of salting-out. According to Lilich and Chernykh, this effect is caused by the redistribution of water among the ions, the greater the energy of the salting-out ion, the more water is fixed and the sharper is the solubility decrease.

(ii) Type II. The salt solubility increases monotonically (the salting-in effect). The reasons for this are the formation of complexes, the hydrolysis of the electrolyte added, the increased ability of the electrolyte to destructure water and, hence, to increase its solving ability. The authors suppose that in all such cases 'free' water is released.

(iii) Type III. The salt solubility passes through the maximum (which is typical of poorly soluble salts). The idea of a solution three-zone structure is applied here; that is, around an ion there is a strictly arranged layer of water molecules, then there is a destructured water layer and, finally, there is a pure water structure. With high electrolyte concentrations, the salting-in effect occurs due to the increasing water volume present in the destructured layers. With high electrolyte concentrations the salting-out effect takes place when all the water is contained in the first hydrate layers.

(iv) Type IV. The salt solubility passes through the minimum. This can be explained either by the fact that, in contrast to Type II, the formation of the complexes begins only above a certain electrolyte concentration or, if the absence of the complexes is proved, the effect is due to the redistribution of water. For example, increasing the concentration of an acid added as a second electrolyte leads to a decreasing dissociation of this acid and, hence, to the release of water.

For such considerations the notions of the positive and negative hydration of ions developed in numerous works by Krestov (see, for example, Krestov and Abrosimov 1967, 1972) are essential. As the temperature increases the water structure becomes disordered, therefore at low temperatures the destruction of the water structure by ions prevails over its ordering due to formation of hydrate spheres of ions (i.e. negative hydration), and at high temperatures the ions do not enhance disorder, since water is already disordered and the ordering of the water structure is observed (i.e. positive hydration). The temperature above which negative hydration is replaced by positive hydration rises with increasing cation ionic radius; for example, it is $\simeq 11^\circ C$ for sodium and $\simeq 89^\circ C$ for caesium. Bivalent and trivalent ions are positive at all hydration temperatures. Therefore, in conclusion, it is reasonable to expect the salting-out effect of ions to increase when negative hydration is replaced by positive hydration.

The experimental data represented above can be completely accounted for qualitatively in similar ways for cases where the absence of solid solutions is ascertained. Since no direct experiments have provided evidence for the presence of solid solutions, the following is another attempt to examine whether the behaviour of the solubility isotherms can give any information on the presence or absence of solid phases in the systems being discussed.

The above considerations were mostly qualitative, since no quantitative theory of solutions is available at present. This being the case a thermodynamic approach to the problem seems to be reasonable. Therefore the approach to be used is the general method of the deduction of phase diagrams proceeding from relations between the thermodynamic potentials of the phases.

Assuming the equations for the thermodynamic potential surface of a liquid phase and the thermodynamic potential of solid phases to be known, then the equations for the solubility isotherms of these phases can be deduced. Here the essential question is to what extent the shape of the experimentally determined isotherms can define the characteristic properties of the thermodynamic potential of the phases, i.e. the properties of these phases.

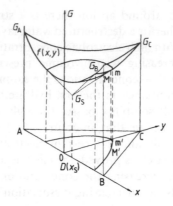

Figure 1.40 The construction of the solubility isotherm in a ternary system.

Choosing the orthogonal coordinate system as shown in figure 1.40, then the origin of the coordinates will be placed in the middle of side AB of the equilateral triangle representing the concentrations. The triangle height will be taken to be unity and the component concentrations will be related to the coordinates x and y by the following

$$[A] = (1 - y - x\sqrt{3})/2 \qquad [B] = (1 - y + x\sqrt{3})/2 \qquad [C] = y.$$

$$(1.28)$$

The component C refers to the solvent (water). Point D in figure 1.40 corresponds to the chemical combination of components A and B, its coordinates being $y = 0$, $x = x_s$.

Let the equation of thermodynamic potential of liquid be given as

$$G = f(x, y) \qquad (1.29)$$

then, since the liquid phase is a homogeneous solution, the surface G should be convex at all points. The basic curvature radii, R, of the surface defined by equation (1.29) are calculated as the roots of the quadratic equation

$$\Delta R^2 + H\delta R + H^4 = 0$$

where

$$\Delta = f_{xx}f_{yy} - f_{xy}^2$$
$$\delta = 2f_x f_y f_{xy} - (1 + f_x^2)f_{yy} - (1 + f_y^2)f_{xx} \qquad (1.30)$$
$$H = (1 + f_x^2 + f_y^2)^{1/2} > 0.$$

Here the indices designate the corresponding partial derivatives of function (1.29). The requirement of the surface G being convex relative to the plane xy is equivalent to the condition that two curvature radii should be positive.

Using the known properties of the roots of quadratic equations we find that it is necessary that conditions $\Delta > 0$ and $\delta < 0$ should be satisfied. Since f_x and f_y may have opposite signs, then following from equation (1.30) the above requirements are satisfied only when f_{xx} and f_{yy} are positive at all surface points.

If at some points of the surface G, $\Delta = 0$, then at these points one of the basic curvature radii is equal to infinity and the basic normal section is a straight line. The case where at $\Delta = 0$ the basic normal section has an inflection point is excluded because otherwise a part of the surface would lie below the tangent plane, i.e. the solution would be limited.

The equation for straight lines that are at a tangent to the surface G at the point (x, y, G) and pass through the thermodynamic potential figurative point $(x_s, 0, G_s)$ of the solid phase, has the form

$$G = G_s - (x_s - x)f_x + yf_y. \tag{1.31}$$

The equation of the solid phase solubility isotherm which is a projection of the tangent points on the xy plane will be obtained in an implicit form by substituting equation (1.29) into (1.31) giving

$$f + (x_s - x)f_x - yf_y - G_s = 0. \tag{1.32}$$

If two solid phases can coexist at a given temperature and the figurative points of their thermodynamic potentials are $(x_{s1}, 0, G_{s1})$ and $(x_{s2}, 0, G_{s2})$, then the invariant point composition corresponding to the intersection of two isotherms will be found by jointly solving two equations of the form of (1.32). Subtracting one equation from the other it is found that the relation between both coordinates of the invariant point (x_i, y_i) is

$$f_x\Big|_{\substack{x = x_i \\ y = y_i}} = \frac{G_{s1} - G_{s2}}{x_{s1} - x_{s2}}. \tag{1.33}$$

In the case of the formation of solid solutions, their thermodynamic potential will be characterized not by the point $(x_s, 0, G_s)$ but by the curve $G_s(x_s)$ which lies in the plane $y = 0$ and is convex relative to the x axis at all its points. Drawing a set of tangent lines relative to G from each point of the curve $G_s(x_s)$ and projecting the tangent points onto plane xy, we obtain a series of curves, each curve being described by equation (1.32). The envelope of this series of curves will be the solubility isotherm of solid solutions. In order to obtain its equation we must find the derivative of equation (1.32) with respect to x_s in order to eliminate x_s. Differentiating (1.32) with respect to x_s gives

$$f_x = dG_s(x_s)/dx_s. \tag{1.34}$$

Hence, $x_s = \varphi(f_x)$, where φ is a certain function whose form is defined by the form of $G_s(x_s)$. Substituting x_s into (1.32) we obtain the isotherm equation for solid solutions:

$$f + [\varphi(f_x) - x]f_x - yf_y - G_s(\varphi(f_x)) = 0. \tag{1.32a}$$

In order to find the coordinates of the isotherm extreme points we shall differentiate the isotherm equation according to the rules of differentiating implicit functions. This gives (for equation (1.32a))

$$f_x + f_y y' + (\varphi - x)(f_{xx} + f_{xy}y') + [\varphi'(f_{xx} + f_{xy}y') - 1]f_x$$
$$- y(f_{xy} + f_{yy}y') - f_y y' - G'_s \varphi'(f_{xx} + f_{xy}y) = 0.$$

Reducing this and taking (1.34) into account gives

$$\frac{dy}{dx} = -\frac{(\varphi - x)f_{xx} - yf_{xy}}{- yf_{yy} + (\varphi - x)f_{xy}} \tag{1.35a}$$

and similarly for equation (1.32) we obtain

$$\frac{dy}{dx} = -\frac{(x_s - x)f_{xx} - yf_{xy}}{- yf_{yy} + (x_s - x)f_{xy}}. \tag{1.35}$$

The coordinates of the isotherm extreme are found by taking dy/dx to equal zero and solving the equation obtained jointly with the isotherm equation.

Now suppose there is a non-singular extremum with coordinates x_e, y_e on the isotherm (1.32). Then at this point, the numerator of equation (1.35) equals zero whilst the denominator does not. Hence, the relation connecting the coordinates of the extremum points is given by

$$(x_s - x_e)/y_e = f_{xy}/f_{xx} \tag{1.36}$$

where the second-order derivatives are taken at the corresponding point on the liquid thermodynamic potential surface and the condition $\Delta \neq 0$ is satisfied (otherwise the denominator of (1.35) will also equal zero and the corresponding point on the isotherm will be singular).

Note that if the isotherm extremum lies on the straight line joining points C and D of figure 1.40 and if the concentration relation of components A and B is constant, then it follows from equation (1.28) that

$$(x_s - x_e)/y_e = x_s = f_{xy}/f_{xx}.$$

In the general case the latter relation should not necessarily be satisfied. Moreover, the extremum position on the isotherm does not refer to any characteristic features of the surface G. In particular, in the section of G produced by a plane $y = y_e$ the condition $f_x|_{y=y_e} = 0$ corresponds to the minimum of the curve obtained. This condition is entirely independent of equation (1.36) and thus the extremum on the isotherm m' should not correspond to the thermodynamic potential minimum of liquid M (figure 1.40). The latter would not in any way affect the behaviour of the isotherm.

In the case where on the surface G there are singular points corresponding to the intersection of its two slopes in the singular secant plane (corresponding to a small dissociation of D in the solution), a singular point will appear on the isotherm. In fact, given this case, at these singular points of G both the first and second partial derivatives will be indefinite, therefore relation (1.35) will be indefinite as well. Indeed, the nodal point on the isotherm may indicate not only the singularity at the corresponding point of G but also that the basic curvature radius is equal to infinity (the numerator and denominator both being equal to zero in equation (1.35) results in the first of the relations in equation (1.30) being equal to zero also).

The above considerations can be illustrated in a form easy to grasp. Let us draw a cylindrical surface through the isotherm, the generating lines being perpendicular to the concentration triangle plane. This surface will intersect that of the liquid phase thermodynamic potential; however, we shall have no information about the intersection line apart from the fact that its projection would produce the initial isotherm. The liquid thermodynamic potential surface is independent of whether the isotherm is a section of a syncline or anticline and whether its branches are linear or not. As it has already been noted the only information is provided by the isotherm singular points.

Thus, the experimental data on the isotherm behaviour do not enable us to obtain information concerning the appearance of the surface G; however, they can help to verify the assumption of a functional relation $G = f(x, y)$ since the problem of calculating the isotherm with a known function $f(x, y)$ is solved in a unique way.

To illustrate, consider a case when the liquid phase is an ideal solution. The thermodynamic potential of this solution is written as

$$G = [A](\mu_A + \varepsilon \ln[A]) + [B](\mu_B + \varepsilon \ln[B]) + [C](\mu_C + \varepsilon \ln[C])$$

$$(1.37)$$

where $\varepsilon = RT$, R is the gas constant, the concentrations are in mole fractions, and μ_A, μ_B, μ_C are the thermodynamic potentials of the initial components of the liquid.

Substituting equation (1.28) for the concentrations in (1.37) and calculating the partial derivatives gives the isotherm equation:

$$2G_s = \mu_A(1 - x_s\sqrt{3}) + \mu_B(1 + x_s\sqrt{3}) + \varepsilon x_s\sqrt{3}\ln\left(\frac{1 - y + x\sqrt{3}}{1 - y - x\sqrt{3}}\right)$$

$$+ \varepsilon \ln\left(\frac{(1 - y)^2 - 3x^2}{4}\right).$$

$$(1.38)$$

In the particular case with $x_s = -1/\sqrt{3}$, i.e. for the solubility of a pure component A, we obtain from (1.38)

$$0.5(1 - y - x\sqrt{3}) = \exp[-(\mu_A - G_s)/\varepsilon]. \qquad (1.38a)$$

The left-hand side of this equation is the concentration of A in the solution, which must be constant since it follows from (1.38a). Thus, the solubility isotherm of A is a straight line parallel to side BC of the concentration triangle. Similarly, with $x_s = 1/\sqrt{3}$ the solubility isotherms of B will be straight lines parallel to side AC.

With $x_s = 0$ the isotherm has the form

$$(1 - y)^2 - 3x^2 = 4\exp[(2G_s - \mu_A - \mu_B)/\varepsilon] \qquad (1.38b)$$

i.e. it is a hyperbola with a vertex at point $x_e = 0$,

$$y_e = 1 - 2\exp[(2G_s - \mu_A - \mu_B)/2\varepsilon] = 1 - \exp[-(G_s^l - G_s)/\varepsilon]$$

where G_s^l is the thermodynamic potential of a liquid having the same composition as the solid phase. The hyperbola asymptotes are the sides of the concentration triangle.

These results can easily be seen in a more general and somewhat more easily understood conclusion. Rewriting equation (1.37) in the form

$$G \equiv f = \sum_{i=1}^{n} a_i(\mu_i + \varepsilon \ln a_i)$$

where a_i is the concentration of the ith component, it is found that

$$f_x = \sum_{i=1}^{n} [a_{ix}(\mu_i + \varepsilon \ln a_i) + \varepsilon a_{ix}]$$

$$f_y = \sum_{i=1}^{n} [a_{iy}(\mu_i + \varepsilon \ln a_i) + \varepsilon a_{iy}]$$

$$G_s = \sum_{i=1}^{n} \{(\mu_i + \varepsilon \ln a_i)[a_i + (x_s - x)a_{ix} - ya_{iy}] + \varepsilon[(x_s - x)a_{ix} - ya_{iy}]\}.$$

Since $\Sigma a_i = 1$, $\Sigma a_{ix} = \Sigma a_{iy} = 0$, and the second term after the summation sign becomes zero. Besides, if a_i are the linear function of the coordinates, then $a_i - xa_{ix} - ya_{iy} = p_i$ and, finally, we obtain

$$G_s = \sum_{i=1}^{n} [(p_i + x_s a_{ix})(\mu_i + \varepsilon \ln a_i)].$$

In the particular case with $p_i + x_s a_{ix} = 1$, which corresponds to the coordinate of a pure component a_i, we have $G_s = \mu_i + \varepsilon \ln a_i$. With $x_s = 0$ and considering $p_1 = p_2 = p$, and $p_3 = 0$ we have

$$G_s = p(\mu_1 + \mu_2 + \varepsilon \ln a_1 a_2).$$

At any point on the isotherm of equation (1.38) the first derivative has the form

$$dy/dx = 3[x_s - x/(1 - y)]/[1 + 3x_s x/(1 - y)]. \qquad (1.39)$$

The maximum on the isotherm (the solubility minimum) refers to the relation $x_s = x_e/(1 - y_e)$ and therefore it always lies on the straight line passing through compositions of both a solvent and a solid phase. Note that with $y = y_e$ the surface minimum (see equation (1.38)) defined from condition $f_x = 0$ takes place with the following coordinate relation

$$x/(1 - y) = (k - 1)/\sqrt{3}(k + 1) \qquad k = \exp[(\mu_A - \mu_B)/\varepsilon].$$

In the general case the right-hand side of this relation is not equal to x_s and, hence, the minimum G does not coincide with the isotherm extremum. (In a two-component system (i.e. $y = 0$), $(k - 1)/\sqrt{3}(k + 1) = x_m$, where x_m is the minimum coordinate of the ideal liquid solution thermodynamic potential.)

As one can see from the above discussion of perfect solutions in a binary system, these considerations are valid only for the solubility isotherms of a chemical compound completely dissociated into its initial components in solution. If a compound is dissociated incompletely, the solution of its components cannot be considered ideal.

Turning now to the case of the formation of binary solid solutions (dissociating appreciably in a liquid phase). For the sake of simplicity let us assume that G_s is the concentration quadratic function

$$G_s = q(x_s - x_m)^2. \qquad (1.40)$$

For a limited range of solid solutions this assumption is always possible if the homogeneity region is small. According to equation (1.34), x_s will be expressed by f_x and substituting it into (1.35a) we obtain the conditions determining the extremum coordinates

$$x_m\sqrt{3} + (3/4q)[\mu_B - \mu_A + \varepsilon \ln\{[1 + x_e\sqrt{3}/(1 - y_e)]/$$
$$[1 - x_e\sqrt{3}/(1 - y_e)]\}] = x_e\sqrt{3}/(1 - y_e). \qquad (1.41)$$

It follows from (1.41) that the isotherm extremum lies on the straight line passing through the composition of the solid phase thermodynamic potential minimum and the solvent composition, but this is true only in the case where $(1 - y_e)$ is equal to zero, i.e. only in the particular case when

$$\frac{1 + x_m\sqrt{3}}{1 - x_m\sqrt{3}} = \exp\frac{\mu_A - \mu_B}{\varepsilon}.$$

This corresponds to the condition that x_m should refer to the thermodynamic potential minimum not only of a solid but also to that of an ideal liquid phase in a binary system AB. Otherwise the isotherm extremum is outside the given straight line.

Let us summarize the results obtained from the consideration of the solubility isotherm equations of a binary compound in a ternary system:

(i) The isotherm non-singular extremum lies on the straight line joining the compositions of the solvent and the solid phase only in the case where

    (*a*) the liquid phase is an ideal solution, i.e. its thermodymamic potential is described by equation (1.37), the solid phase being dissociated completely into its initial components; or

    (*b*) the compound composition is identical to the composition referring to the thermodynamic potential minimum of an ideal binary liquid solution. If the compound is a solid solution, its composition should also be identical to the composition referring to the thermodynamic potential minimum of the solid solution.

(ii) The isotherm singular extremum always lies on the above-mentioned straight line and indicates the absence of solid solutions. In this case, the thermodynamic potential surface of the liquid has two slopes which intersect on the singular secant plane. The solid phase dissociates weakly in solution into its initial components.

(iii) The proof of the existence of solid solutions is the absence of the isotherm singular points lying on the above-mentioned straight line. The presence of a non-singular extremum, in itself and regardless of its location, does not indicate the existence of solid solutions.

Our analysis has essentially led to the same conclusions derived by M I Ravich (1938, 1940) 50 years ago, although some additional information has been added (see (i)).

In view of the discussion presented above let us consider the characteristic features of the state diagrams studied.

On most isotherms analysed there is a singular minimum corresponding to a stoichiometric solution. As a consequence, the thermodynamic potential surface has two intersecting slopes. This fact should correspond to the weak dissociation of mono-substituted orthophosphates into their initial components such as an ultra-acid salt and a doubly substituted ortho-phosphate. This fact does not imply weak electrolytic dissociation, which is in fact rather high in acid salts, but it means that the solution compositions in acid and alkali regions of the system differ greatly in ionic composition (due to a significant difference between the first and second dissociation constants of the orthophosphoric acid). Indeed, when a mono-substituted orthophosphate is dissolved in water or an acid medium, it is only H_2PO_4 anions that are produced, the content of HPO_4 ions in the solution being very small. Even the addition of a small amount of alkali to the aqueous solution sharply increases the number of these ions.

The explanation for the distinct linearity of the isotherm branches, which is particularly the case for the acid branch, requires special assumptions to be made concerning the type of dependence of the solution thermodynamic potential. We have shown that a linear behaviour of the isotherms (the solubility constancy) is observed for components producing ideal solutions. It can be assumed that this is the case here. So it is best not to consider the alkali–acid–water systems, where a violent chemical reaction proceeds, but rather systems of the types

$$A(H,D)_2BO_4 . m_1 A_2(H,D)BO_4 - A(H,D)_2BO_4 - A(H,D)_2BO_4 . nH_2O$$

$$A(H,D)_2BO_4 . m_2 (H,D)_3BO_4 - A(H,D)_2BO_4 - A(H,D)_2BO_4 . nH_2O$$

where the relation n/m_i is taken so that the concentration triangle side should be parallel to the isotherm. In such systems no chemical reaction will take place and our assumption would be self-consistent. The smoothing-out of the solubility minimum with increasing temperature can now be treated as a decrease in the isotherm slope with respect to the x axis due to a decreasing ratio n/m which becomes possible because of the difference in the stability of the complexes as the temperature rises. One can of course consider other types of dependence $G = f(x, y)$ resulting in linear isotherms (for example, by making the isotherm parallel to itself in such a way that its vertex would describe a curve convex with respect to the y axis in the plane $x = 0$). However, such a dependence would require additional assumptions for its justification.

The disappearance of the singularity and the shift of the solubility isotherm minimum may be caused both by the formation of solid solutions and by increasing the salting-out effect of cations as their atomic number increases. Since D_3PO_4 is a weaker acid in comparison with H_3PO_4, the salting-out effect of the alkali cations in deuterated solutions may also be stronger. This problem can be resolved only by experiments. It is clear, however, that if solid solutions do exist, their homogeneity region is rather narrow.

1.7 Isotope effects

Three characteristic features distinguish the behaviour of mono-substituted orthophosphates, and the orthophosphates of alkali metals and ammonium in deuterated aqueous solutions from that of many other compounds.

First, there is the much greater solubility in D_2O of the compounds in question as compared to the solubility in H_2O. Second, there is the changing structure of the crystallizing solid phase with the increasing deuteration degree of solution. Third, there is the fact that the distribution coefficient of deuterium in solid and liquid phases differs considerably from unity.

Below we shall discuss questions concerning the hydrogen isotope disproportionation and then the solubility isotope effect is considered.

1.7.1   Hydrogen isotope exchange equilibrium

The influence of the solution deuteration degree, x_1, on the deuteration degree of crystals, x_s, in equilibrium with the solution has not been studied extensively. It is known that the deuteration degree of Na_2SO_4.10aq and $SrCl_2$.6aq does not differ from the deuteration degree of solutions from which they have been grown (Rabinovich 1968). At the same time the deuteration degree of mono-substituted potassium orthophosphate is much less than that of the solution (Havránková and Březina 1974, Loiacono et al 1974, Belouet et al 1975, Belouet 1980 and Bespalov et al 1977). Similar observations were made with mono-substituted caesium orthophosphate (Balascio et al 1975).

Our study of KB_5O_8.$4(D_xH_{1-x})_2O$ showed that the deuteration degree of solutions (pH $\simeq$ 8.0; $0.4 < x < 0.96$) did not differ from that of crystals. The same result was obtained with solutions of $LiCOOH.(D_xH_{1-x})_2O$ (pH $\simeq$ 7.2; $0.48 < x < 1$). It should be noted that in the latter case the isotope exchange occurred in a crystallization water molecule and a proton bound with carbon was not replaced by a deuteron (Shevchik et al 1977). We also showed that the deuteration degree of mono-substituted orthophosphates of potassium, rubidium and caesium and the orthoarsenate of caesium is less than that of the corresponding saturated solutions (Momtaz and Rashkovich 1976).

To determine the deuteration degree of solutions and solid phases, the spectrophotometric analysis of the water released by the thermal decomposition of crystals or by solution evaporation is usually used. This method was suggested by Lemeshko and Egorov (1971) and was first applied by Dmitrenko and Korolikhin (1975) in their study of $K(H,D)_2PO_4$ crystals. Later the author and co-workers improved the method (Volkova et al 1975) which was then applied to other substances as well (Shevchik et al 1977). The developed technique allows the determination of the deuteration degree to an accuracy of $\simeq$ 0.005 with $x > 0.3$, and to a slightly lower accuracy with lower values of x.

To determine the deuteration degree of a solution, complete thermal dehydration of the precipitated solid phase was carried out after evaporation of the solvent. Therefore the deuteration degree measured characterized the solution as a whole. Figure 1.41 gives a sketch of the device used to extract water from the substances being studied. All the experiments using heavy water and deuterated solutions were carried out in an airtight transparent box, with dry nitrogen blown through it. The results of our experiments are given in table 1.19. The table shows that in all cases x_s is much less than x_1.

Figure 1.42 represents all the available data on the isotope equilibrium between solution and $K(D_xH_{1-x})_2PO_4$ crystals; the data were obtained at various temperatures and in different conditions. One can see that the data

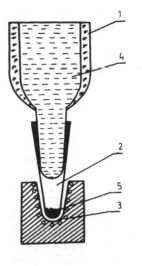

Figure 1.41 Sketch of the device used for the thermal decomposition of salts with the simultaneous freezing of water vapour: 1, Dewar; 2, test tube; 3, heater; 4, liquid nitrogen; and 5, sample.

are in fairly good agreement with our results. The only previously known experimental result on $Cs(D_xH_{1-x})_2AsO_4$ crystals is also given in table 1.19; however, the agreement with our results in this case is much worse.

The data in table 1.19 were processed using the least-squares method by three empirical formulae

$$x_s = x_1 \exp[-b(1 - x_1)] \qquad (1.42)$$

$$x_s = x_1\alpha/[1 - x_1(1 - \alpha)] \qquad (1.43)$$

$$x_s = x_1 \frac{x_1 + (1 - x_1)k_{22}}{x_1^2 + 2x_1(1 - x_1)k_{22} + (1 - x_1)^2 k_{11}k_{22}}. \qquad (1.44)$$

The calculated constants of these equations are given in table 1.20 along with the mean-square error in the approximation of σ_0. For all cases the value of σ_0 is roughly the same and of the same order as the accuracy of the deuteration degree determination. Slightly lower values of σ_0 for equation (1.44) can easily be accounted for: in contrast to other formulae, this one contains not one but two empirical constants.

In non-deuterated (with $x = 0$) and completely deuterated (with $x = 1$) solutions the relation $x_1 = x_s$ should be true. Formula (1.42) (Loiacono *et al* 1974) was the first to be used among various functional relations that satisfy this condition. However, it seems that Loiacono and co-workers did not deduce any physical meaning from this formula.

Table 1.19 Isotope exchange equilibrium in mono-substituted ortho-phosphate solutions of potassium, rubidium and caesium and in caesium orthoarsenate at 25°C.

Orthophosphates						Orthoarsenate	
K (pH ≃ 4.3)		Rb (pH ≃ 4.2)		Cs (pH ≃ 4.0)		Cs (pH ≃ 4.9)	
x_1	x_s	x_1	x_s	x_1	x_s	x_1	x_s
Tetragonal solid phase				Monoclinic solid phase		Tetragonal solid phase	
0.156	0.108	0.175	0.112	0.423	0.352	0.334	0.258
0.360	0.264	0.318	0.228	0.488	0.419	0.384	0.308
0.376	0.266	0.436	0.348	0.516	0.442	0.420	0.334
0.544	0.430	0.471	0.364	0.546	0.471	0.477	0.384
0.644	0.536	0.518	0.408	0.585	0.520	0.551	0.470
0.692	0.594	0.578	0.480	0.617	0.556	0.674	0.595
0.726	0.632	0.638	0.564	0.659	0.597	0.692	0.620
0.760	0.675	0.690	0.602	0.694	0.640	0.930[a]	0.856[a]
0.820	0.740	0.742	0.674	0.724	0.669		
0.854	0.788			0.753	0.701		
0.980	0.960						
Monoclinic solid phase		Solid phase II					
0.832	0.790	0.858	0.838				
0.912	0.882	0.884	0.865				
0.950	0.935	0.960	0.948				

[a] Data from Balascio et al (1975) for a solution of pH = 8.6 at 40°C.

Processing their data on $K(D_xH_{1-x})_2PO_4$ saturated solutions, Bespalov et al (1977) used the expression

$$x_s = x_1/[k - x_1(k - 1)]$$

which is a particular case of equation (1.44) with $k_{11} = k_{22} = k$. With $k = \alpha^{-1}$, the expression is transformed into equation (1.43). Bespalov and co-workers obtained $k = 1.65$ ($\alpha = 0.606$) which is close to our values of $k_{11} = 1.60$, $k_{22} = 1.54$ and $\alpha = 0.637$. Let us discuss the physical meaning of the constants involved in equations (1.43) and (1.44).

Since the discovery of deuterium in 1932 and the synthesis of heavy hydrogen water in 1933, hydrogen–deuterium isotope exchange reactions have been studied extensively. The results obtained were generalized in a number of review works, that of Varshavskii and Vaisberg (1957) deserving special attention. Most works are devoted to isotope exchange in a homogeneous medium (mainly in solution). Less information is available on

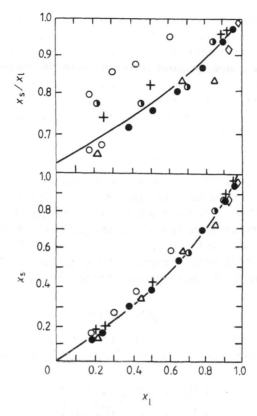

Figure 1.42 Deuterium distribution between $K(H,D)_2PO_4$ crystals and saturated solution. The curves are plotted according to the author's data using equation (1.44). ○, 30°C; △, 45°C; ◑, 60°C (Havránková and Březina 1974); ◇, Belouet (1980); +, Loiacono *et al* (1974); and ●, Bespalov *et al* (1977).

Table 1.20 Constants of the approximating equations (1.42)–(1.44).

Formula of the salt	(1.42)		(1.43)		(1.44)		
	b	σ_0	α	σ_0	k_{11}	k_{22}	σ_0
$K(H,D)_2PO_4$	0.5155	0.0049	0.6375	0.0049	1.604	1.539	0.0049
$K(H,D)_2PO_4$[a]	0.3260	0.0034	0.7500	0.0032	1.040	1.399	0.0034
$Rb(H,D)_2PO_4$	0.4171	0.0149	0.6887	0.0167	1.784	1.180	0.0096
$Cs(H,D)_2PO_4$	0.2958	0.0043	0.7610	0.0049	1.420	1.226	0.0034
$Cs(H,D)_2AsO_4$	0.3784	0.0046	0.7070	0.0054	1.482	1.334	0.0047

[a] Monoclinic solid phase.

the isobar equilibrium between liquid and gas, whilst the heterogenic equilibrium between liquids and solids has not been studied to any extent.

Let us consider reactions involving isotope exchange between components AI_n and BI_m in a homogeneous medium using the results obtained by Varshavskii and Vaisberg (1957). Let us assume that element I has isotopes I^1 and I^2, and that A and B are parts of molecules without atoms of element I. Between these molecules some exchange reactions can proceed, each of which can be written as

$$AI^1_{n-i+1}I^2_{i-1} + BI_{m-j}I^2_j \rightleftharpoons AI^1_{n-i}I^2_i + BI^1_{m-j+1}I^2_{j-1} \qquad (1.45)$$

where $i = 1, 2, \ldots, n$; $j = 1, 2, \ldots, m$. Altogether, nm reactions of this type can take place. In the particular case where $A \equiv B$ formula (1.45) can also be applied; however, $n = m$ and therefore the number of reactions would be less than n^2, since expressions with $i = j$ become equalities, and the reaction pairs with $i = k$, $j = l$ and $i = l$, $j = k$ prove to be identical. One can easily see that in this case the number of reactions would be $n(n-1)/2$. For example, it is possible to write only one exchange reaction between the water hydrogen isotopes

$$H_2O + D_2O \rightleftharpoons 2HDO \qquad (1.46)$$

$A = B = O$, $I^1 = H$, $I^2 = D$, $n = m = 2$, $i = 1$, $j = 2$.

Equilibrium constants for reactions of type (1.45) are determined by the formula

$$k_{ij} = \frac{[AI^1_{n-i}I^2_i][BI^1_{m-j+1}I^2_{j-1}]}{[AI^1_{n-i+1}I^2_{i-1}][BI^1_{m-j}I^2_j]}. \qquad (1.47)$$

If isotopes I^1 and I^2 are distributed uniformly (equiprobably) among both components, the equilibrium constant can easily be calculated. Let the relative concentration in the system of isotope I^2 equal x. For the case of uniform distribution, the concentration of this isotope in each component will be the same. The numerical value of x shows the probability of atom I being atom I^2 in any molecule. The probability of any molecule in the system being, for instance, $AI^1_{n-i}I^2_i$ molecule is equal to $C^i_n x^i (1-x)^{n-i}$, where C^i_n is the number of combinations of n elements taken i at a time. This probability is numerically equal to the concentration of the given molecules. Similarly,

$$[BI^1_{m-j}I^2_j] = C^j_m x^j (1-x)^{m-j}. \qquad (1.48)$$

Substituting formula (1.48) into (1.47) gives

$$k_{ij\infty} = \frac{C^i_n C^{j-1}_m}{C^{i-1}_n C^j_m} = \frac{(n-i+1)j}{(m-j+1)i} \qquad (1.49)$$

where $k_{ij\infty}$ is the equilibrium constant of equation (1.45) with an equiprobable isotope distribution. For the particular case of reaction (1.46) the equilibrium constant is found to be 4.

Given the equilibrium constant, it is not difficult to calculate the concentrations of the equilibrium isotope complexes (molecules or ions) participating in reaction (1.46). Solving the set of equations

$$[H_2O] + [D_2O] + [HDO] = 1$$

$$[D_2O] + 0.5[HDO] = x \qquad (1.50)$$

$$k = \frac{[HDO]^2}{[H_2O][D_2O]}$$

gives

$$[HDO] = \{[1 + 4Fx(1-x)]^{1/2} - 1\}/F$$

$$[D_2O] = x - 0.5[HDO] \qquad (1.51)$$

$$[H_2O] = 1 - x - 0.5[HDO]$$

where $F = (4 - k)/k$.

With an equiprobable isotope distribution when $k = 4$, it follows from (1.48) that

$$[HDO] = 2x(1-x) \qquad [D_2O] = x^2 \qquad [H_2O] = (1-x)^2. \qquad (1.52)$$

The experimental value determined for k for reaction (1.46) is 3.80 at 25°C (Kirshenbaum 1951). As the temperature rises, $k \Rightarrow k_\infty$. In fact, at 25°C the calculation using (1.51) differs from the approximate calculations using equation (1.52) by a mere 1.2%.

With $A = B$ and $n = m > 2$ the condition of equiprobable isotope distribution allows the calculation of the concentrations of equilibrium complexes using formula (1.48). For example, in the case involving isotope exchange between H_3PO_4 and D_3PO_4 ($n = 3$) the concentrations would be obtained by including the expression $[x + (1-x)]^3$:

$$[D_3PO_4] = x^3$$

$$[D_3HPO_4] = 3x^2(1-x)$$

$$[DH_2PO_4] = 3x(1-x)^2$$

$$[H_3PO_4] = (1-x)^3.$$

It should also be noted that for reactions involving isotopic exchange between AI_n^1 and BI_m^2 the summary reaction can be written

$$mAI_n^1 + nBI_m^2 \rightleftharpoons mAI_n^2 + nBI_m^1 \qquad (1.53)$$

where the equilibrium constant is

$$k = \prod_{i=1}^{n} \prod_{j=1}^{m} k_{ij}. \qquad (1.54)$$

One can see (using equation (1.49)) that with an equiprobable isotope distribution

$$k_\infty = \prod_{i=1}^{n} \prod_{j=1}^{m} k_{ij\infty} = 1.$$ (1.55)

It is not sufficient to know the equilibrium constant of reaction (1.53) in order to calculate the concentrations of all the equilibrium isotope complexes, their number being $n + 1$ for the first component, and $m + 1$ for the second one. The total number of unknowns is $n + m + 2$ and in order to find them the same number of equations is required. Writing the concentration balance between both components and the summary content of isotope I^2 gives three equations, the remaining $n + m - 1$ equations can be provided by expressions for any k_{ij}, their number (nm) being more than enough. Therefore it is only $n + m - 1$ equilibrium constants that need to be determined by experiment; the other constants can be determined after the calculation of the unknown values of the equilibrium concentrations.

When $A = B, (n + 1)$ isotope complex concentrations are unknown. When the concentration of component A remains constant and the known summary concentration of one isotope yields two equations, the remaining ones can be written if the experimentally determined $(n - 1)$ equilibrium constant is chosen from $n(n - 1)/2$ constants.

It is usually quite difficult to determine the concentrations of separate isotope complexes by experiment. To determine the value of the isotope distribution coefficient of two components or phases is much easier. The distribution coefficient can be determined by different methods, but the two most common methods are used here. Let x be the deuteration degree, and indices 1 and 2 refer to the two components or phases, then

$$\alpha = [x_1/(1 - x_1)]/[x_2/(1 - x_2)]$$ (1.56)

$$K = x_1/x_2.$$ (1.57)

Equation (1.56) is taken as the relative deuterium distribution coefficient and (1.57) determines the distribution coefficient. Thus, the relative distribution coefficient (α) indicates how many times the ratio of the deuterium and hydrogen concentrations in one component (or phase) is greater (or less) that in the other; the distribution coefficient (K) characterizes the ratio of the deuteration degrees of the components (or phases). The relative distribution coefficient usually refers not to a separate exchange reaction (e.g. equation (1.45)) but to the summary exchange.

The connection between the relative distribution coefficient and the equilibrium constant of the isotope exchange reaction was found by Varshavskii and Vaisberg (1957) with certain allowances. They showed that

$$\alpha = k_{ij}/k_{ij\infty} = \text{constant}$$ (1.58)

i.e. α does not depend on the deuteration degree. Using equations (1.54) and (1.55) gives the following expression for the summary reaction (1.53):

$$\alpha = (k/k_\infty)^{1/nm}$$

In fact, equation (1.58) is equivalent to the assumption of an equiprobable distribution of isotopes between isotopic forms of one component. The validity of this assumption has been proved by experimental data showing that the equilibrium constants of an isotope exchange in a one-component homogeneous system are close to those referring to the equiprobable isotope distribution. Deviations decrease with temperature, and do not usually exceed 5%. It is only for the reactions $H_2 + D_2 \rightleftharpoons 2HD$ at 25°C that $k = 3.26$ and $k_\infty = 4$ (Kirshenbaum 1951).

Thus, given the relative distribution coefficient, the equilibrium constant can be calculated. Note that, following from equation (1.58), the relative distribution coefficient is always equal to unity within the equiprobable isotope distribution regime.

Transforming expression (1.56) for the relative distribution coefficient to the form

$$x_1 = x_2\alpha/[1 - x_2(1 - \alpha)] \tag{1.59}$$

one can see that it coincides with empirical formula (1.43). It has been shown (table 1.20) that the experimental data can be described by this dependence, i.e. the value of α can be considered to be independent of the solution deuteration degree. If follows that the equiprobable isotope distribution holds for water, for the dissolved salt and for the crystal (taken separately). The question of whether the relative distribution coefficient refers to the exchange reaction in solution or to the isotope exchange between the solid phase and the solution or to both processes is left open.

Havránková and Březina (1974) followed by Bespalov et al (1977) studied the deuterium redistribution between crystals and solution of $K(H,D)_2PO_4$. Thye believed that the ratio of the isotopic salt forms entering the solid phase is the same as that in solution, i.e. the value of α characterizes the isotope exchange in solution. Baikov (1986) was of the same opinion and believed that hydrogen isotopes in water and acid anions are not equivalent thermodynamically and their distribution, therefore, differs from the equiprobable one. These concepts give rise to the following objections.

Firstly, α differs too much from unity, while for all other solutions α differs from unity by not more than a few per cent. Secondly, the value of α is shown to depend on the structure of the solid phase of the same composition. Thirdly, all the salts under investigation are dissociated into ions to a considerable extent, and in different salts the isotope exchange proceeds between the same ions of phosphoric acid and water, therefore the strong effect of the cation on the value of α cannot be understood. Fourthly, the deuteration degree of a liquid and its vapour are usually different. It is common knowledge that the relation of these values is approximately equal to that of the vapour pressures above liquids containing different isotopes. Similarly, one can expect the relation of the deuteration degree of the dissolved salt and crystal to differ from unity.

The total relative distribution coefficient is equal to the product of the values of α for the exchange between water and dissolved salt and between the dissolved salt and the crystal. To determine the former is a quite difficult experimental task. Proceeding from this we assumed that the deuterium disproportionation occurs mainly during crystallization, therefore $\alpha \simeq 1$ for the isotope exchange reactions between water and phosphoric acid ions. In this case the deuterium degrees of water, dissolved salt and solution as a whole must be the same and the distribution of isotopes in solution should be equiprobable. The relative concentration of isotopic forms would be defined by expression (1.52); similar expressions can also be written for the isotopic forms of the phosphoric acid ions. Thus it is believed that the deuterium disproportionation occurs not in solution but during crystallization. Let us consider this problem in more detail.

The isotope exchange between water and the crystal surface is described by reaction (1.45) with the equilibrium constants of equation (1.47). Writing the corresponding equations in the explicit form (where suffix s refers to the solid phase, the cation is not specified for simplicity of notation, and B represents either P or As):

$$i = 1, j = 1 \qquad H_2O + (HDBO_4)_s \rightleftharpoons HDO + (H_2BO_4)_s \qquad (1.60)$$

$$i = 1, j = 2 \qquad H_2O + (D_2BO_4)_s \rightleftharpoons HDO + (HDBO_4)_s \qquad (1.61)$$

$$i = 2, j = 1 \qquad HDO + (HDBO_4)_s \rightleftharpoons D_2O + (H_2BO_4)_s \qquad (1.62)$$

$$i = 2, j = 2 \qquad HDO + (D_2BO_4)_s \rightleftharpoons D_2O + (HDBO_4)_s. \qquad (1.63)$$

Similar equations can also be written for the isotope exchange reactions between the dissolved salt anions and the crystal. Their equilibrium constants, however, are uniquely connected with the equilibrium constants of reactions (1.60)–(1.63) and to similar reactions between water and the dissolved salt anions, which can readily be calculated by taking (1.49) into account and assuming an equiprobable isotope distribution between the solution components.

Having written the equations for the constants of equilibrium for (1.60) and (1.63) we obtain the following relations

$$\frac{[HDBO_4]_s}{[H_2BO_4]_s} = \frac{[HDO]}{[H_2O]} \frac{1}{k_{11}} \qquad \frac{[HDBO_4]_s}{[D_2BO_4]_s} = \frac{[HDO]}{[D_2O]} k_{22} \qquad (1.64)$$

$$\frac{[H_2BO_4]_s}{[D_2BO_4]_s} = \frac{[H_2O]}{[D_2O]} k_{11} k_{22}.$$

Multiplying the first two equalities of (1.64) we have

$$k_s = \frac{[HDBO_4^2]_s}{[H_2BO_4]_s[D_2BO_4]_s} = \frac{[HDO]^2}{[H_2O][D_2O]}\frac{k_{22}}{k_{11}} = 4\frac{k_{22}}{k_{11}} \qquad (1.65)$$

where k_s is the equilibrium constant of the hypothetical reaction

$$(H_2BO_4)_s + (D_2BO_4)_s \rightleftharpoons 2(HDBO_4)_s.$$

This reaction does not, of course, take place in the crystal, but it characterizes the equilibrium between isotope complexes on the crystal surface arising during crystallization and the coexistence with the solution.

Relation (1.65) can also be obtained from the equilibrium constants of reactions (1.61) and (1.62):

$$k_s = k_{22}/k_{21} = 4k_{22}/k_{11} = k_{12}/4k_{21} \qquad k_{12}/k_{22} = 4. \qquad (1.66)$$

From the condition

$$[H_2BO_4]_s + [D_2BO_4]_s + [HDBO_4]_s = 1$$

of equalities (1.52) and (1.64) we obtain

$$[D_2BO_4]_s = x_1^2/[x_1^2 + 2x_1(1 - x_1)k_{22} + (1 - x_1)^2 k_{11}k_{22}]$$

$$[HDBO_4]_s = 2x_1(1 - x_1)k_{22}/[x_1^2 + 2x_1(1 - x_1)k_{22} + (1 - x_1)^2 k_{11}k_{22}]$$

$$(1.67)$$

$$[H_2BO_4]_s = (1 - x_1)^2 k_{11}k_{22}/[x_1^2 + 2x_1(1 - x_1)k_{22} + (1 - x_1)^2 k_{11}k_{22}].$$

Assuming the surface deuteration degree and the solid phase volume to be the same, $x_s = [D_2BO_4]_s + 0.5[HDBO_4]_s$ can now be written as

$$x_s = x_1[x_1 + (1 - x_1)k_{22}]/[x_1^2 + 2x_1(1 - x_1)k_{22} + (1 - x_1)^2 k_{11}k_{22}].$$

$$(1.44)$$

Dividing both parts of (1.44) by x_1 gives the expression for the distribution coefficient K and so the relative distribution coefficient (see equation (1.56)) can now be written in the form

$$\alpha = \frac{1}{k_{22}}\frac{x_1 + (1 - x_1)k_{22}}{x_1 + (1 - x_1)k_{11}}. \qquad (1.68)$$

One can see from equations (1.44) and (1.68) that K and α depend on x_1. The dependence of α on the deuteration degree results from the non-equiprobable isotope distribution in the solid phase. In this case the relative distribution coefficient has no advantages, the K value being easier and more convenient in practical applications.

Note that with the equiprobable isotope distribution in the crystal $k_s = 4$, it follows from equation (1.65) that $k_{11} = k_{22} = k$, i.e. $\alpha = k^{-1} = $ constant. In this case, equation (1.44) is transformed into (1.43) and (1.59).

It should be noted that this is the case only if x_1 and x_s do not change during crystallization. This is in fact the case with the compounds under study, since the solubility of the salts does not exceed a few mol %, the content of hydrogen and deuterium in the solid phase being about 2 mol %. Thus, for instance, if a 1 kg crystal is grown from a 10 kg solution of mono-substituted potassium orthophosphate with an initial deuteration degree of 90%, then by using the temperature reduction technique (with an initial temperature of 50°C) by the final stage of the process (at 29.6°C) $x_1 = 90.5\%$. The difference in the crystal deuteration degree from the initial to the final stages of the growing process is about 0.2%.

The experiments described above show that formula (1.46) gives $k_s = 4k_{22}/k_{11} \neq 4$. Deviation of k_s from $k_s = 4$ is much greater than the experimental and calculational errors. Thus it is concluded that the distribution of protons and deuterons in the crystal differs from the equiprobable one. The relative concentrations of complexes at the crystal surface can be calculated using formulae (1.51). Figure 1.43 represents the relative concentrations at different values of k_s.

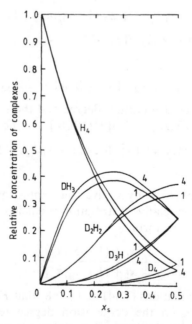

Figure 1.43 Relative concentrations of $AD_iH_{2-i}BO_4$ complexes on a crystal surface. k_s values are designated at the curves. The curves for i-complexes with $x_s > 0.5$ and those of $(2 - i)$-complexes with $x_s < 0.5$ are symmetrical.

It should be noted that the value of k_s is defined only by the solid phase properties and the effect of the crystallization conditions cannot be significant. Indeed, changing the solution pH results in increasing the content of H and H_2BO_4 ions in the solution with decreasing pH, and increasing the OH and HBO_4 ions with increasing pH. Since the isotope exchange reactions in the solution are characterized by an equiprobable distribution of isotopes and the solid phase properties are independent of the solution pH, the isotope exchange itself between the solution and crystal cannot vary in any appreciable way. The emergence of HBO_4 ions does not change the situation, because the equilibrium constants of their isotope exchange with the crystal would be uniquely connected with the isotope exchange reactions of these ions with the solution and with the known equilibrium constants. The temperature effect on the isotope exchange in the solution is revealed in the fact that with increasing temperature all equilibrium constants approach the ones calculated assuming the equiprobable distribution. The solid phase properties in the temperature region usually employed for crystal growth (20–70°C) do not practically change. Thus the temperature effect upon the distribution coefficient and on k_s cannot be significant. These conclusions have been confirmed by experiment (Momtaz and Rashkovich 1976).

It is reasonable to ask why the deuterium distribution coefficient in the systems under study is different from unity and why the relative content of hydrogen in the crystal is higher than that in the solution. It should be remembered that in crystals of other structural types, where hydrogen bonds are not of importance, deuterium disproportionation between the crystal and saturated solution does not occur.

It is reasonable to believe that if complexes containing protons are more likely to enter the crystal, their binding energy is higher. The higher energy of the hydrogen bond (as compared with the deuteron one) should apparently be attributed to the possibility of proton tunnelling between its equilibrium positions in the two-minimum potential well. This is confirmed by the decreasing temperature of high-temperature phase transitions of $K(H,D)_2PO_4$ and $Rb(H,D)_2PO_4$ crystals as their deuteration degree increases. These phase transitions are due to the disruption of hydrogen and deuterium bonds, therefore the higher the deuteration degree, the lower the temperature at which the phase transition occurs.

Note that the deuteron bond with its 'own' oxygen is stronger than that of the proton. Accordingly, the ferroelectric transition temperature in the crystals under study increases with increasing deuteration degree. The fact is that the transition is connected with the proton/deuteron sublattice disordering which requires the proton (deuteron) to 'break' from its 'own' oxygen. A stronger deuteron bond with its 'own' oxygen results in a higher lattice energy of the deuterated compounds.

The values

$$\varepsilon_D = \frac{[D_2BO_4]_s/[D_2BO_4]_l}{[H_2BO_4]_s/[H_2BO_4]_l} = \frac{1}{k_{11}k_{22}} \qquad \varepsilon_{HD} = \frac{[HDBO_4]_s/[HDBO_4]_l}{[H_2BO_4]_s/[H_2BO_4]_l}$$

$$= \frac{1}{k_{11}} \qquad (1.69)$$

can be considered quantitatively as the ratio of the binding energies at the crystal surfaces of D_2BO_4 and $HDBO_4$ complexes to the binding energy of the H_2BO_4 complex. This ratio is proportional to the ratio of probabilities of the solution entering crystals of the corresponding complexes. The constants of equation (1.69) given in table 1.21 were calculated using the experimental data represented in table 1.20. Table 1.21 shows that $\varepsilon_D \simeq 0.5$ and increases slightly with the increasing atomic number of the cation.

Each BO_4 tetrahedron (B being P or As) in the structure of the KH_2PO_4 family crystals is connected by hydrogen bonds with four other tetrahedrons and is therefore surrounded by four hydrogen isotopes. Therefore $H_{2-i}D_iBO_4$ complexes cannot be said to exist inside the crystals. Following the ferroelectric cluster model (Slater 1941), $D_jH_{4-j}BO_4$ clusters are usually considered to exist in the case of a uniform distribution of hydrogen between BO_4 groups, their relative concentration b_j being described by formula (1.48):

$$b_{j\infty} = \frac{4!}{j!(4-j)!}x_s^j(1-x_s)^{4-j} \qquad (j = 0\text{--}4). \qquad (1.70)$$

When calculating the values of b_j, the initial assumption is that at the crystal surface the protons and deuterons cannot migrate between the oxygen atoms of the neighbouring BO_4 groups since this migration would give rise to charged configurations because some oxygen atoms at the surface would have no neighbours to exchange their protons (deuterons) with. So at the surface we have a relative concentration a_i of $D_iH_{2-i}BO_4$ groups ($i = 0\text{--}2$). It is these values that were calculated above (formulae (1.67)). Given the value of k_s, these values can also be calculated by equation (1.51).

The second assumption is as follows. When $(H,D)_2BO_4$ ions or complexes containing this ion become attached to the crystal, hydrogen (deuteron)

Table 1.21 Constants characterizing the hydrogen isotope exchange between the crystal surface and solution.

Salt	k_s	ε_D	ε_{HD}	$(4-k_s)/4$ (%)
$K(H,D)_2PO_4$	3.84 ± 0.14	0.41 ± 0.02	0.62 ± 0.02	4.0 ± 3.5
$Rb(H,D)_2PO_4$	2.65 ± 0.22	0.48 ± 0.05	0.56 ± 0.03	34 ± 6
$Cs(H,D)_2PO_4$	3.45 ± 0.11	0.57 ± 0.01	0.70 ± 0.02	14 ± 3
$Cs(H,D)_2AsO_4$	3.60 ± 0.13	0.51 ± 0.03	0.67 ± 0.02	10 ± 3

bonds between the ion and the corresponding surface site are established. As a result the proton (deuteron) of an ion becomes bound with the surface oxygen, or the proton (deuteron) emerging on the surface becomes bound with the oxygen ion. If the hydrogen isotope long-range interaction between the crystal and the isotopes entering the crystal complex is not taken into account, then the latter can attach to any site at the crystal surface. In other words, we believe that the probability of the attaching complex being $D_i H_{2-i} BO_4$ is proportional to the concentration of such complexes at the surface and does not depend on the neighbouring isotope complex. The deuteration degrees of the surface and the bulk crystal are assumed to be the same.

Simple, though somewhat inelegant, calculations give

$$
\begin{aligned}
b_4 &= a_2 x_s^2 \\
b_3 &= a_1 x_s^2 + a_2 2 x_s (1 - x_s) \\
b_2 &= a_0 x_s^2 + a_1 2 x_s (1 - x_s) + a_2 (1 - x_s)^2 \\
b_1 &= \qquad\quad a_0 2 x_s (1 - x_s) + a_1 (1 - x_s)^2 \\
b_0 &= \qquad\qquad\qquad\qquad\quad a_0 (1 - x_s)^2 .
\end{aligned}
\tag{1.71}
$$

Figure 1.44 represents the relative concentration calculated from relation

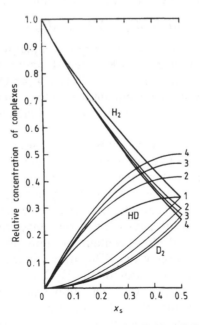

Figure 1.44 Relative concentrations of $AD_j H_{4-j} BO_4$ complexes in a crystal structure with different values of k_s (designated at the curves).

(1.71) with two values of k_s. Note that it follows from the equalities

$$a_0 + a_1 + a_2 = 1$$

$$a_2 + a_1/2 = x_s$$

that the value of a_2 with $x = x_s$ equals the value of a_0 with $x = 1 - x_s$. Hence, as one can see in (1.71), the curves in figure 1.44 are symmetric with respect to the straight line $x_s = 1/2$. Therefore the curves are plotted with x_s ranging from 0 to 0.5.

The deviation of the hydrogen isotope distribution in the crystal structure from the equiprobable one can be described by $(4 - k_s)/4$ (see the right-hand column in table 1.21). The data show that the value reaches few tens of per cent.

1.7.2 Heavy water as a solvent

Solubility (L) is known to be connected with the dissolution free energy (ΔG is the isobar–isothermic potential variation) by the relation

$$- RT \ln L = \Delta G = \Delta H - T\Delta S \tag{1.72}$$

where ΔH and ΔS are enthalpy and entropy changes, respectively.

It is generally assumed that when salts are dissolved in H_2O and D_2O the entropy term is either small or varies slightly and the dissolution enthalpy plays the dominant role in equation (1.72).

The dissolution enthalpy value can be represented as the sum of three terms

$$\Delta H = \Delta H_c + \Delta H_w + \Delta H_h \tag{1.73}$$

where ΔH_c is the crystalline lattice disruption enthalpy, $\Delta H_c > 0$; ΔH_w is the disruption enthalpy of the near-order water structure, $\Delta H_w > 0$ (this effect is mainly due to the disruption of some $OH\ldots O$ (or $OD\ldots O$) bonds when foreign ions (or molecules) are implanted into the water structure); and ΔH_h is the hydration enthalpy, i.e. the interaction of the dissolved particles with water, $\Delta H_h < 0$.

In most cases $\Delta H > 0$ and the solubility increases with temperature, hence, in the right-hand side of equation (1.73) the sum of the first two terms is usually greater than the third one.

Let us consider the change in ΔH when crystals are dissolved in ordinary and heavy water. We shall consider three main cases: (i) the solid phase does not contain protons (deuterons); (ii) the solid phase contains crystallization water; and (iii) the solid phase contains protons (deuterons) that are not involved in crystallization water. The upper index 'H' will denote the corresponding enthalpies of dissolution in H_2O, and index 'D' those in D_2O

In the former case $\Delta H_c^H = \Delta H_c^D$. The dipole moments of H_2O and D_2O being equal (Kirshenbaum 1951), the electrostatic interaction of water with the dissolved ions is the same in both cases: hence, $\Delta H_h^H = \Delta H_h^D$. The disruption energy of deuterium bonds in D_2O is greater than that of hydrogen bonds in H_2O (by $\simeq 330\,cal\,mol^{-1}$). Hence

$$\Delta(\Delta H) = \Delta H^D - \Delta H^H = \Delta H_w^D - \Delta H_w^H > 0.$$

A more positive dissolution enthalpy of these substances in D_2O, as compared with H_2O, has been confirmed by experiments in all cases. Also according to equation (1.72), the solubility in D_2O is less than that in H_2O.

The increase of the ionic radius of the salt cation (or anion) leads to a disruption of the water structure. As a result the salt solubility usually increases and the contribution of ΔH_w into the enthalpy variation increases as well. At the same time, the difference $\Delta H_w^D - \Delta H_w^H$ becomes greater, i.e. the isotope effect is enhanced. On the other hand, elevating the temperature leads to a decrease in isotope effect both in the solubility value and in the dissolution enthalphy. This is caused by the fact that, with rising temperature, the water structure near-order disappears gradually and the distorting effect of foreign ions decreases (the contribution of ΔH_w in equation (1.72) diminishes).

Brodskii (1952) and Rabinovich (1968) compared the solubilities of inorganic substances in ordinary and heavy water. Holmberg (1968) summarized the data on phase equilibria of many sodium and potassium salts with solutions in H_2O and D_2O. Table 1.22 represents some of his data confirming the conclusions given above.

Table 1.22 Solubility isotope effect (L) of salts in ordinary and heavy water (Rabinovich 1968, Holmberg 1968). $\Delta L = (L^H - L^D)/L^H (\%)$ where L is in salt mol per water mol.

Compound	Temperature (°C)	ΔL	Compound	Temperature (°C)	ΔL
KF	40	0.7	Na_2CO_3.aq	30	1.5
KCl	40	7.6	Na_2CO_3.aq	60	3.3
KBr	40	8.9	K_2CO_3.aq	20	1.9
NaCl	40	4.6	K_2CO_3.aq	60	2,7
NaCl	0	7.2	Na_2SO_4.7aq	0	22.2
$NaNO_3$	10	1.5	$CuSO_4$.5aq	40	5.0
KNO_3	10	13.3	$MnSO_4$.aq	0	16.0
Na_2SO_4	50	1.2	$MnSO_4$.aq	25	44.0
K_2SO_4	50	6.1	$BeSO_4$.6aq	25	0.6
K_2SO_4	20	10.0	$CdSO_4$.aq	50	7.2
$KBrO_3$	25	10.5	NaF.2aq	20	0.5
KIO_3	25	25	LiCOOH.aq[a]	25	2.8
$LiIO_3$[a]	25	18.8	LiCOOH.aq[a]	50	1.9
$LiIO_3$[a]	47.5	14.2	KB_5O_8.4aq[a]	24.4	5.1
			KB_5O_8.4aq[a]	44.5	3.8

[a] Shevchik *et al* (1977).

In all cases studied, the solubility of substances that do not contain protons (or deuterons) in D_2O is less than in the cases of H_2O. In particular, it proves the validity of the assumption of the dominant role of the enthalpy term of equation (1.72).

When the crystal contains water of crystallization, it usually splits off during dissolution. Since the enthalpy of mixing H_2O with D_2O is approximately the same as that of D_2O with H_2O, the above considerations remain valid. Experimental measurements of the isotopic effects in this case, however, are rather difficult due to isotope exchange between the crystal surface and solution. Yet comparing the dissolution enthalpy of a deuterated salt in D_2O with that of a non-deuterated salt in H_2O we should point out that $\Delta H_c^D \neq \Delta H_c^H$. Hence

$$\Delta(\Delta H) = \Delta(\Delta H_c) + \Delta(\Delta H_w).$$

The crystalline lattice disruption enthalpy of a deuterated crystal is greater than that of a non-deuterated one, therefore both terms are greater than zero and the solubility in D_2O should be less than that in H_2O. As for crystalline hydrates, whose solubility diminishes with increasing temperature (the absolute value of ΔH_h being quite high and $\Delta H < 0$), increasing ΔH_w during dissolution in D_2O makes ΔH less negative and slows down the decrease of solubility with rising temperature. Therefore the solubility isotope effect may increase with rising temperature. This is illustrated by the right-hand section of table 1.22.

However, there are some crystalline hydrates, whose solubility in H_2O in a certain temperature range is less than that in D_2O (NaBr.5aq, NaF.2aq, $SrCl_2$.6aq, LiCl.aq), although the differences in solubility of these salts in H_2O and D_2O are quite small (Holmberg 1968).

We shall now discuss separately the substances, belonging to group (iii), in which protons (deuterons) pass into sodium during dissociation as weakly hydrated ions or as parts of complex ions forming strong bonds with water. The former include mono-basic acids and mono-acid bases as well as ammonium salts, while the latter include polyatomic acids and bases, and acid salts.

As for the isotope effect, the behaviour of mono-basic acids and bases should not differ appreciably from that of the crystalline hydrates discussed above: for example, our experiments (Volkova et al 1971b) showed that at $20-50°C$ the solubility of DIO_3 in D_2O is about 7 wt % less than that of HIO_3 in H_2O. Ammonium salts appear to behave in the same way. Unfortunately, there are few relevant experimental data on this topic. The only fact available is that the solubility of $(ND_4)_2SO_4$ at $20-60°C$ in D_2O is less than that of $(NH_4)_2SO_4$ in H_2O by $1-2\%$ (Rabinovich 1968).

As far as acid salts are concerned, their anions containing hydrogen isotopes, as a rule, form hydrogen (deuteron) bonds with water molecules. In this case, deuteration results in changing all three terms in equation

(1.73) and it is difficult to predict the sign of the isotopic effect. There are experimental data only for mono-substituted orthophosphates and orthoarsenates of alkali metals and ammonium.

The enthalpy of dissolution of RbH_2PO_4 and RbD_2PO_4 in H_2O and D_2O was studied by the author and co-workers and by A F Vorob'ev and his co-workers (Monaenkova *et al* 1972, 1978, Vorob'ev *et al* 1974). It was established that the formation of deuterium bonds between the dissolved salt and D_2O decreases the enthalpy of the dissolved salt as compared with its enthalpy in H_2O, and this must have resulted in increasing its solubility. However, the crystalline lattice enthalpy (which is more positive in deuterated crystals) makes a greater contribution to the value of the dissolution enthalpy, therefore $\Delta H^D > \Delta H^H$ and, if the entropy factor is not taken into account, the solubility in D_2O should be less than that in H_2O.

Let us now consider the above data on the solubility of substances under study in H_2O and D_2O. The results are represented in table 1.23.

The table shows that in all cases an 'anomalous' isotopic effect on the solubility is observed, which is much greater than the value of ΔL for other substances (see table 1.22). The effect decreases with rising temperature (except for mono-substituted caesium orthoarsenate) and with the increasing atomic weight (and ionic radius) of the cation.

Table 1.24 illustrates the solubility isotopic effect for acid and alkaline solutions, giving data on the solubility isotherms for potassium systems (see table 1.16) since other salts behave in a similar way. One can see that the solubility in deuterated solutions is higher, but with increasing concentration of an acid or alkali, the isotopic effect decreases; the decrease in an alkaline medium being much greater.

In order to correlate these data with the calorimetric measurements described previously one should take into account the entropy term in the

Table 1.23 Isotope effect on the solubility in water of mono-substituted orthophosphates and orthoarsenates of alkali metals and ammonium (L is in moles per 55.51 moles water, $\Delta L = (L^H - L^D)/L^H (\%)$).

Salt	25°C			50°C		
	L^H	L^D	ΔL	L^H	L^D	ΔL
$N(H,D)_4(H,D)_2PO_4$	3.13	4.90	−56.8	5.17	7.75	−49.9
$K(H,D)_2PO_4$	1.85	2.94	−59.4	3.01	4.44	−47.6
$Rb(H,D)_2PO_4$	4.25	6.17[a]	−45.1	6.61	8.44[b]	−27.7
$Cs(H,D)_2PO_4$	6.18	6.81	−10.1	8.44	8.51	−0.8
$Cs(H,D)_2AsO_4$	9.2	14.5	−58	11.8[c]	20.4[c]	−73[c]

[a] Phase II.
[b] Phase I.
[c] Values at 55°C.

Table 1.24 Isotope effect on the solubility of $K(H,D)_2PO_4$ in acid and alkaline solutions at 25°C. (Concentrations are in moles per 55.51 moles of water.)

Concentration of K_2O, P_2O_5 in solvents	L^H	L^D	$L^H - L^D$	$L^H - L^D/L^H$ (%)
		Acid medium		
1	2.83	3.93	−1.10	−39
2	3.84	5.07	−1.23	−32
4	5.86	7.35	−1.49	−25
		Alkaline medium		
1	4.08	5.08	−1.0	−24
2	6.35	7.17	−0.82	−13
4	10.90	11.34	−0.44	−4

expression for the dissolution free energy. The variation of the system entropy involved in dissolving the substances in question in D_2O must be considerably greater than for the case of H_2O.

As the data show, the absolute entropy of deuterated compounds is always higher. Therefore one can assume that the entropy of both the initial components (salt and water) and the solution increases with the deuteration degree. Various experimental data can be used to show how the entropy changes during dissolution.

Comparing data on solutions in H_2O and D_2O, for example, one can see that, although the molecular volumes of the initial H_2O and D_2O, and RbH_2PO_4 and $Rb(H,D)_2PO_4$ are practically the same, their partial molecular volumes in solutions are different; the difference being greater in D_2O than in H_2O. Thus, at 25°C the difference in partial molar volumes of RbH_2PO_4 and H_2O is $54.1 - 18.0 = 36.1$ cm^3, while that in the deuterated solution is $56.7 - 17.6 = 39.1$ cm^3 (see table 1.10). This increased difference in the partial molecular volumes of the components should be associated with a greater entropy variation during dissolution.

The entropy term in the expression for the dissolution free energy in deuterated solutions is greater because the difference in the interaction energy of uniform solution particles is greater, which results in a greater deviation from the equiprobable distribution. Indeed, as has already been mentioned, at constant temperature the maximum specific conductivity of $NH_4H_2PO_4$ and RbH_2PO_4 solutions is observed at a lower concentration than for similar solutions in heavy water. This fact indicates that the interaction of the dissolved salt particles between each other is more intense if they do not contain deuterium. Also the interaction of water molecules is more vigorous in D_2O than that in H_2O. Thus, the difference in the interaction energy between D_2O molecules and that between deuterated salt particles is greater than that of similar pairs in ordinary water.

The weaker interaction between particles of deuterated rubidium salt as compared with the hydrogenated salt can be accounted for by its lower compression during dissolution. The data given in table 1.10 indicate that the molecular volume of RbH_2PO_4 during dissolution decreases by 10.0 cm^3 and that of RbD_2PO_4 by 7.8 cm^3 (at $25°C$).

Thus, the above considerations concerning the entropy factor of the anomalous isotope solubility effect agree with all available data on the properties of these salt solutions. The greater variation of the entropy term in the expression for the dissolution free energy in deuterated solutions appears to be connected with a greater disordering of the salt structure rather than any change in the water, since even with a high salt concentration ($\simeq 10$ water molecules per one salt molecule) the partial water density differs only slightly from the pure water density, while the partial density of the salt in solution is much higher than the density of the crystals.

2 Mechanism and Kinetics of Crystallization

In what follows it will be shown that KH_2PO_4-family crystals grow by the dislocation mechanism. A screw dislocation emerging onto a face forms a non-vanishing step. Crystallizing matter attaching to the step twists it into a spiral. As a result a dislocation hillock appears and spreads out across the face and at the normal to it thereby causing the crystal to grow. Our aim is to discuss the peculiarities of this mechanism.

The rate of crystallization as for any other process is determined by its slowest stage. In crystal growth there are assumed to be two major stages: supplying crystallizing matter from solution to the crystallization surface and the formation of a crystal. If it is the first stage that is the limiting step, then the growth is said to proceed in the diffusion regime; if it is the second stage which is the limiting step, then the growth regime is said to be kinetic. The intermediate case is called the mixed-kinetics regime. It is clear that in order to study the surface processes it is necessary to perform experiments in the kinetic regime; its characteristic feature is the independence of the growth rate to the solution stirring intensity, i.e. the flow velocity of the solution flowing around the crystal. However, this condition is not usually realized in practical crystal growing.

Most of this chapter is based on the experiments carried out in the kinetic regime. The first section is devoted to the interference technique of studying characteristics of crystal growth *in situ* perfected by the author and co-workers. The next four sections mainly deal with the peculiarities of the mechanism and kinetics of dislocation growth. In the last section the effect of the solution hydrodynamics and that of other factors on the morphological stability of the growing surface is discussed.

In this chapter the generally accepted notation for crystals in physics and technology is adopted: ADP for $NH_4H_2PO_4$, DADP for $ND_4D_2PO_4$, KDP for KH_2PO_4, and DKDP for KD_2PO_4.

2.1 Research technique

In the study of crystal growth kinetics the major property to be measured has usually been the normal growth rate of the face, R, determined by the face shift using an optical microscope. Typically the accuracy of such measurements has been $\simeq 100\ \mu$m, the highest accuracy attainable being $\simeq 3\ \mu$m. Interference methods to determine R allow the shift of the faces to be measured to an accuracy of $\simeq 0.2\ \mu$m and, more importantly, they enable the automation the measurement process (Bobylev and Rashkovich 1980, Rashkovich *et al* 1982, 1983).

The surface relief details can be studied comparatively easily after the crystal has been removed from solution. It is only in the last decade that it became possible to study the relief details during the crystallization process. A detailed review of the methods used is given in the work of van Enckevort (1984). Today, growth layers 14 Å high can be revealed *in situ* on the surface of potassium biphthalate crystals (Tsukamoto 1983), growth layers 80 Å high being revealed on potassium dihydrogen phosphate crystals (Dam and van Enckevort 1984). The methods developed together with the *in situ* x-ray topography of the crystal dislocation structure (Chernov *et al* 1979) made it possible to obtain previously inaccessible information on the crystallization details. Nevertheless, the quantitative aspect of these phenomena is not fully clear.

Although the technique used by us does not allow elementary steps to be distinguished, it does provide detailed information concerning the location and shape of dislocation hillocks on the surface of a growing crystal and allows the normal (R) and tangential (v) growth rate to be measured at any point of the face relief (Rashkovich *et al* 1985, Rashkovich and Shustin 1987).

The optical layout of the experimental arrangement is given in figure 2.1; this is a Michelson interferometer. The light beam of the laser (1) is broadened up to $\simeq 1$ cm in diameter and is divided by the half-transparent mirror (or beam-splitting cube) (4). The objective (9) produces an image of the growing face of the crystal (7) in the plane of the camera (11) and in the plane of the detector window of the television camera (12). The interference fringes arising as a result of the interference of the light beams reflected from the crystal and the reference mirror (6) are superposed on this image. The filter (5) is used to match the intensities of both light beams. One or several photo detectors can be attached at any point of the crystal surface image on the television screen (13). The intensity variation of the incident light can be detected by the recorder (15). The microscope (16) is then used to determine the image scale.

The crystal is placed into the flow-through cell; the solution can be made to flow either along or perpendicular to the growing face. During the growth (dissolution) of the crystal, the solution layer thickness between the cell window and the growing face varies thereby causing a variation in the phase

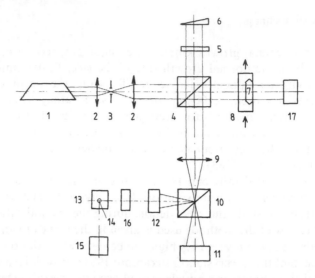

Figure 2.1 Optical scheme to determine crystal growth characteristics in solution: 1, laser; 2, telescope; 3, diaphragm; 4 and 10, beam-splitting cubes; 5, filter; 6, reference mirror; 7, crystal; 8, flow-through cell; 9, objective; 11, camera; 12, television camera; 13, television set; 14, photodetector; 15, recorder; 16, video; 17, microscope.

difference of the interfering light beams and a shift of the interference fringes; in other words, a periodic change (with period T) in the illumination intensity at a given point in the interference pattern is produced.

If the face is flat then straight parallel interference fringes (the distance between which depends on the angle between the object and the reference light beams) will be superposed on the image of the face. This angle can be adjusted by turning both the cell containing the growing crystal and the reference mirror. The adjustment is hindered by the low intensity of the light reflected from the crystal; this intensity being less, the smaller the difference between the refractive indices of the crystal and the solution. However, even working with a laser of low power ($\simeq 10 \, \text{mW}$) this adjustment can easily be performed.

In the case when the face contains irregularities, the interference fringes will be curved. After uniform illumination of the flat sections by turning the reference mirror or the cell, the relief can be judged from the shape of the fringes: the interference fringes are equivalent to horizontal contours on a topographic chart. The distance along the relief height between neighbouring fringes equals $\Delta h = \lambda/2n$, where λ is the light wavelength, and n is the solution refractive index. For a helium–neon laser, $\lambda = 0.6328 \, \mu\text{m}$, with typical values of $n \simeq 1.4$ and $h = 0.23 \, \mu\text{m}$. By measuring the distance between the fringes, Δd, (taking the scale into account) the slope of the relief at a given point can be determined using $p = \tan \alpha = \Delta h / \Delta d$, where α is the angle between the line tangent to the surface and the flat section.

It could arise, however, that the face does not contain a flat section corresponding to a crystallographic plane with simple indices. Now p can be determined at random. For example, let the surface to be a right circular cone. Then, by turning the reference beam with respect to the cone axis, one can observe a change in the shape of the interference fringes. A set of concentric rings is now substituted by a set of ellipses, then hyperbolas and parabolas would appear. Note that in this case it is impossible to determine the angle between the cone axis and the crystallographic plane, but the shape of the hillock and its vertex angle can be determined in a unique way.

Nevertheless, the face location can be revealed and the object light beam can be set perpendicular to it. A suitable adjustment is based on the appreciable change of the hillock slope with changing solution supersaturation. Since the growth layers propagate parallel to the face, it is necessary to make the points, where the slope changes, lie in one interference fringe. Such an adjustment makes it possible to determine the shape of the hillocks and to relate p and the crystallographic plane position.

The normal growth rate at any point of the surface is determined from the relation $R = \Delta h / T$. For example, with $T = 1h$, $R \simeq 0.005$ mm per day. The tangential growth rate v can be determined either from the geometric relation $v = R/p$ or directly: when the object beam is normal to the face, and the reference and object beams are parallel, v coincides with the velocity of the interference fringes in the image plane.

The accuracy in measuring p is $\simeq 10\%$ but can easily be increased by photometric measurement of negatives and by using a comparator. The relative accuracy with which T (or R) can be measured from the graph recorded on the automatic plotter is 2–3%.

In figures 2.2 and 2.3 the stages of the regeneration process of the prismatic faces of ADP (figure 2.2) and KDP (figure 2.3) crystals illustrate the potential of the technique described. Comparing the face image in one beam (when the reference mirror is closed) with the same image with the interference fringes superposed upon it, one can see that the latter contains much more information. Thus, the terraces that appear to be plane in figure 2.2 turn out to be hillocks of the elliptic shape. The interference pattern makes it possible to reveal many relief details in figure 2.3.

The solution is supplied into the cell from the airtight thermostat of about 1.5 litres in volume, made of Plexiglas. Stirring is performed with a propeller stirrer. In the thermostat there is a pump ensuring a constant velocity of $\simeq 0.5{-}150$ cm s^{-1} of the solution flowing through the cell. In the case of tangential flow of the solution around the cell, the flow direction can quickly be reversed. Stabilization of the temperature (to $\pm 0.005°C$) is performed by an electronic circuit in such a way that the solution temperature in the cell

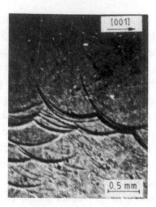

Figure 2.2 Growth layers on the ADP prismatic face. On the right can be seen the photograph of the same surface without the interference pattern.

does not depend on its velocity. The temperature is measured by a resistance thermometer and can be fixed by an automatic recorder, if necessary.

The seed crystal of $5 \times 5 \times 2$ mm is glued to a special holder so that the glue would not hinder the growth of the neighbouring faces. Cells of different designs were employed. The solution layer between the face and the cell glass window ranged from 2 to 10 mm. The window was made at an angle to the face in order to swing the light reflected from it out of the way.

The salts used to prepare the solutions contained less than 1 ppm of impurities of any (the salts being analysed for 13 impurity elements) bivalent or trivalent metals. Unfortunately, the analysis for other ions and possible organic impurities was not carried out. Double distilled water and D_2O with $x = 99.9\%$ and $\chi = 10^{-6} \Omega^{-1}$ cm^{-1} were used.

At the end of this chapter the experimental determination of the concentration of the layer above the surface and the visualization of the solution flows are described. The concentration distribution was judged by the bending of the interference fringes produced in the Mach–Zehnder interferometer (figure 2.4). The cell was placed into one arm of the interferometer in such a way that the light passed parallel to the face under investigation and perpendicular to the solution flow. The objective (7) produced the image of the face edge. In a homogeneous solution the interference fringes were equidistant straight lines. These lines were parallel to the crest of the wedge formed by the wave fronts of the reference and object light beams; the greater the angle of this wedge, the smaller the line spacing. Adjustments (by turning the prisms (5)) were made to ensure that the fringes were perpendicular to the face. Near the growing crystal, where the concentration and the solution refractive index associated with the concentration were less than at a greater distance from the surface (at low velocities of the solution flow), the

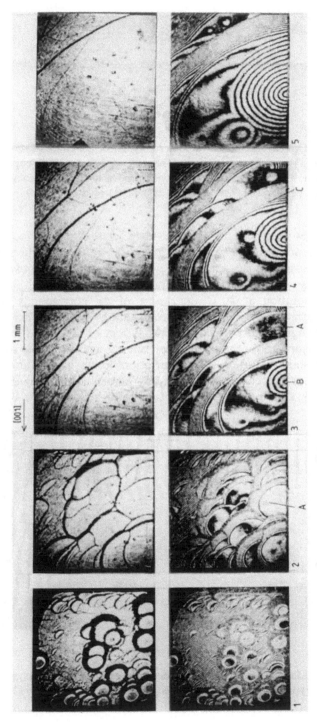

Figure 2.3 Regeneration of the KDP prismatic face. The upper photographs were taken with the reference mirror covered; $t \simeq 30°C$ and $\sigma = 0.022$. Time (in hours): 1, 0; 2, 1; 3, 3.5; 4, 4.5; 5, 6.5. A, plain area; B, dislocation hillock apex; C, macrostep slope.

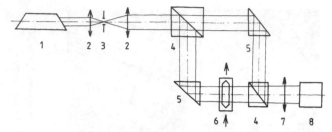

Figure 2.4 Optical scheme for the determination of the concentration distribution in the undersurface layer of the solution: 1, laser; 2, telescope; 3, diaphragm; 4, beam-splitting cube; 5, turning prisms; 6, flow-through cell with a growing crystal; 7, objective; 8, camera or television camera.

interference fringes were bent towards the wedge crest. With the face width l and the interference fringe being shifted by a distance equal to the spacing between neighbouring fringes, the average variation of the refractive index above the face is $\Delta n = \lambda/l$. With $l = 5$ mm and $\lambda = 0.6328\,\mu$m, $\Delta n \simeq 1.3 \times 10^{-5}$. For the ADP solution at $40°C$ this value of Δn corresponds to the concentration variation $\Delta c = 0.1$ wt % (see equation (1.7), section 1.1.4).

A typical photograph taken using such a system as illustrated in figure 2.4 is shown in figure 2.5. In the photograph one can see clearly the solution layer where the major variation in concentration occurs. The solution flow velocity is so small that convection can easily be determined: the depleted solution rises upwards near the lateral faces. One can also see that the

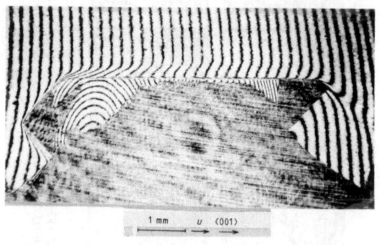

Figure 2.5 Distribution of the concentration of the solution under the ADP crystal surface. The solution flow velocity $u = 0.4$ cm s^{-1}, $\sigma = 0.02$.

concentration near the front (relative to the flow) edge of the prism is less than that above the rest of the face under investigation. This is due to the fact that the dipyramid faces grow, in this case, much faster (by about 10 times) than the prismatic faces do, and the depleted solution above the pyramid drifts onto the prism. Note that the minimum thickness of the solution sublayer which could be determined from the photographs was $\simeq 10 \, \mu m$.

Measurements using the system of figure 2.4 were usually accompanied by observations of the surface relief (using the system detailed in figure 2.1) in the same cell and with the same crystal, which made it possible to correlate the growth rate with the actual supersaturation at a given surface point.

The solution flow near the crystal surface was visualized using bubbles of hydrogen released during the solution electrolysis and carried away by the solution flow. Two platinum wires (of diameter 0.3 mm) were used as electrodes and were attached to the cell in front of and behind the crystal. The photographs are shown in section 3.3.

In conclusion let us consider the calculation of the solution supersaturation, σ, which is the driving force of the crystallization process. By definition

$$\sigma = \Delta\mu/k_B T = \ln(a/a_0) \simeq \ln(c/c_0) \simeq (c - c_0)/c_0. \qquad (2.1)$$

Here $\Delta\mu$ is the difference of the chemical potentials of the solution and the crystal, k_B is the Boltzmann constant, T is the absolute temperature, a and a_0 are the activity of the supersaturated and saturated solutions, respectively, and c and c_0 are the corresponding concentrations. The last approximate equality is valid with $\sigma \ll 1$.

It is necessary to use concentrations instead of activities because there are no data available on activity coefficients. This can be justified assuming that the coefficients are close, when c and c_0 differ slightly. The question of units, in which the concentration is to be measured, *is* important. It is correct to use the volume concentration $c^* = cd$. The data presented in Chapter 1 show that the solution density usually depends linearly on the concentration: $d = A + Bc$; for the ADP solutions in H_2O where $B \simeq 0.006$, and for the KDP solutions where $B \simeq 0.008$ g cm^{-3} per %. (In both cases $A \simeq 1$ g cm^{-3}.) Then

$$\sigma^* = (c^* - c_0^*)/c_0^* = \sigma[1 + Bc/(A + Bc_0)].$$

The term in square brackets is close to a value of 1 and depends weakly on σ and temperature. Also, it is more suitable to use σ and σ^*.

It is not reasonable to use the concentration values in grams per gram of H_2O: $c^{**} = c/(1 - c)$, where c is in mass fractions, in order to determine the supersaturation. Indeed,

$$\sigma^{**} = (c^{**} - c_0^{**})/c_0^{**} = \sigma/(1 - c).$$

The denominator in the right-hand side of this relation depends strongly on temperature since $c \simeq c_0$, therefore the temperature dependences of any characteristics determined at constant supersaturation will be quite different in the σ and σ^{**} scales.

In the experiments described in this chapter the solution supersaturation was created due to a lowering of the temperature and calculated as $\sigma = \ln(c/c_0)$. During the experiment c remained constant, since the solution mass exceeded considerably the amount of the crystallized substance. The value of c_0 was determined by the solution saturation starting temperature t_0. The accuracy to which t_0 was determined was $\pm 0.02°C$. This accuracy was achieved by using a microscope to observe the process of growth and the dissolution of the pyramid faces. The relations given in Chapter 1 were used to calculate c_0. The relative accuracy with which σ was determined depended mainly on the accuracy of the determination of the saturation temperature: $\Delta\sigma/\sigma = \Delta t_0/\Delta t = 0.02/\Delta t$, where $\Delta t = t_0 - t$. When the solution was supercooled by $\Delta t = 1°C$, $\Delta\sigma/\sigma$ was of the order of 2%: when the supercooling step was smaller, the error increased.

2.2 Surface morphology of growing faces

2.2.1 Face formation
The surface of a crystal in solution initially scatters light, and the interference pattern can be observed only after a certain period of time. The initial roughness of the surface is caused by mechanical action and surface dissolution, since the crystal is placed in a slightly overheated solution. After the solution has become supersaturated, the growth begins. The slopes of microroughnesses grow by the normal mechanism, their tangential motion produces flat sections that are oriented parallel to the crystallographic plane (the plane of the singular face) and are located at different heights (figure 2.6). If the angle between the initial surface of the crystal and the singular face is too big ($>1°$), only one plane section appears near the edge, which then becomes the nearest one to the singular face (figure 2.7). Later the plane sections form terraces climbing over each other; such terraces are shown in figure 2.8. Each terrace growing in the tangential direction does not grow upwards (frames A–C).

Dislocations emerging onto the rough surface do not reveal themselves until they appear on the plane sections due to the growth of the latter. The transition of the dislocation from the slope to the plane section can be seen in frame D of figure 2.8. After this it begins to generate a dislocation hillock; its development can be seen in frames E–H. One more dislocation has emerged onto the face in frame G and it generates a hillock (frame H). Dislocations are often not inherited from the seed but emerge on the solution inclusion when the neighbouring pyramid face is regenerated; the growth of this face results in an enlargement of the prismatic face and dislocations pass

Figure 2.6 The initial regeneration stage of the ADP prismatic face.

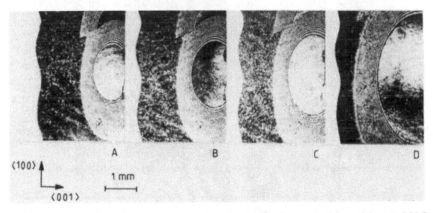

Figure 2.7 Growth of a plane area on the ADP prismatic face with $t \simeq 35°C$, $\sigma = 0.003$ and the time (in hours): A, 0; B, 4; C, 9; D, 28. On the left-hand side the slope is so steep that light does not fall on the objective.

over to it. The formation of the corresponding hillocks is shown in the upper parts of frames E–H (figure 2.8) and in figure 2.9.

It is rather difficult to produce a singular plane face, even if the seed does not contain any dislocations, since in most cases the latter appear on the solution inclusions produced as a result of regeneration (Chernov *et al* 1988).

Face formation does not end after the initial roughness has disappeared and the dislocation hillocks have been formed. These hillocks begin to 'compete' with one another and, if the supersaturation does not change, only one hillock remains; the steepest one to be formed under these conditions wins. When the supersaturation changes, a new hillock generated by another

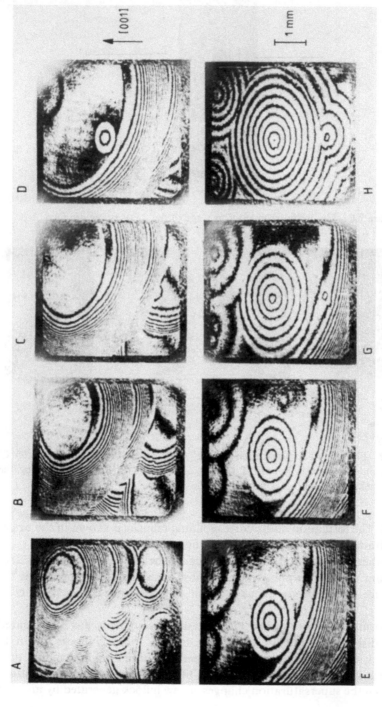

Figure 2.8 Formation of the ADP prismatic face at $t \simeq 35°C$. The experimental parameters are as follows (time τ in minutes, $\sigma \times 10^2$): A, (0, 0.06); B, (21, 0.43); C, (53, 0.43); D, (87, 0.58); E, (107, 0.58); F, (115, 0.58); G, (123, 0.83); H, (141, 1.08).

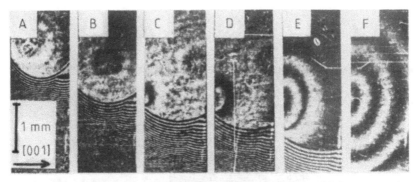

Figure 2.9 Transition of the dislocation from the pyramid face onto the singular area of the ADP prismatic face with $t = 34.7°C$, $\sigma = 0.001$ and the time (in days): A, 0; B, 1; C, 1.9; D, 2.2; E, 3.1; F, 4.4.

dislocation source may appear and persist. This process is illustrated in figure 2.10. Frame A shows the stable relief of the face produced by a hillock whose vertex is located at the lower edge. When the supersaturation increases, two new hillocks appear, one of which turns out to be stable under the altered conditions (frame F). When the supersaturation returns to its initial value, the steepness of the new hillock decreases and the first one reappears, which is steeper at this supersaturation. The face then assumes its former appearance (frame L).

Thus, although many dislocations capable of generating growth hillocks can emerge onto the face, under stationary conditions growth is ensured by only one dislocation source whose hillock occupies the whole face. In principle, the stationary coexistence of two or more hillocks of the same steepness is possible, but this is seldom realized due to reasons to be discussed in section 2.4 in addition to the mechanism causing the 'competition' between the dislocation hillocks.

2.2.2 The shape of dislocation hillocks
Dislocation hillocks on dipyramid faces of ADP and KDP crystals and their deuterated analogues have the shape of trihedral pyramids. A hillock of this

Figure 2.10 Competition of hillocks on the ADP prismatic face. Temperature (°C): A, 34.7; B–D, 33.9; E–F, 33.3; G–L, 34.1.

Figure 2.11 The (101) face of an ADP crystal with one dislocation hillock. Neighbouring vicinal faces are labelled with numbers (Kuznetsov *et al* 1987).

type which occupies the whole face is shown in figure 2.11. Three generating neighbouring vicinal faces are not flat and have different steepness; $p_1:p_2:p_3 \simeq 2.3:1.4:1$. The tilt of each hillock slope along any direction is nearly constant, i.e. its generating lines are straight. The hillock edges do not coincide with the characteristic crystallographic directions; the angles between edges vary slightly for different hillocks and no distinct dependence of these angles on the supersaturation is observed. Figure 2.12 shows the scheme of the arrangement of hillock edges on two pyramid adjacent faces and the corresponding angles are marked. According to the measurements of Hilscher (1985) for KDP: $\alpha = 78°$, $\beta = 115°$ and $\gamma = 167°$. These angles did not vary by more than $\pm 8°$. The bisectors of angles α and β were nearly perpendicular to the corresponding edges: $\delta_1 \simeq \delta_2 \simeq 90°$. On the same vicinal faces the steps are nearly parallel to the edges. On the third vicinal face the orientation of steps was close to the position of the bisector of angle ε. According to the data by van Enckevort *et al* (1980), $\alpha = \beta = \gamma = 120°$ for ADP and KDP and the steps on any vicinal slope move parallel to the pyramid face edges in the ratio $p_1:p_2:p_3 \simeq 6:2:1$. These results seem to be too approximate.

One can see in the figures given in this chapter that hillocks on prismatic faces have an elliptical shape. Torgesen and Jackson (1965) were the first to establish this fact for ADP. At low supersaturations the ratio of half-axes

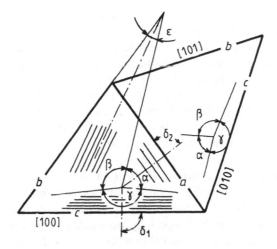

Figure 2.12 Arrangement of hillock edges on the adjacent (101) faces of KDP crystals and the orientation of steps (Hilscher 1985).

for ADP is $\simeq 1.5$ while that for KDP is less than $\simeq 1.35$. The ellipse short axis for ADP is nearly always oriented along [001], but in some cases it deviates from this direction by up to $\simeq 10°$. For KDP the angle between the ellipse short axis and [001] ranges from $\simeq 10$–$22°$. The generating lines of hillocks, just as on the dipyramid faces, are straight lines (Rashkovich *et al* 1985, Chernov *et al* 1987).

With increasing σ the shape of hillocks on prismatic faces changes, this change being reversible and occurring in the narrow range of supersaturations. The successive stages of changing the shape of the KDP prismatic face hillock are shown in figure 2.13. One can see that the ellipse elongates and turns around [001] by $\simeq 45°$ so that its short axis is still at $\simeq 22°$ to [001],

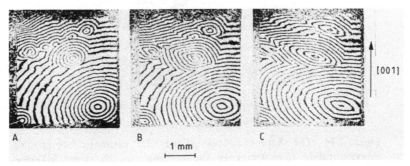

Figure 2.13 Modification of the dislocation hillock on the KDP prismatic face with increasing supersaturation $t = 29°C$. $\sigma \times 10^2$: A, 3.65; B, 3.88; C, 4.00.

but now it is on the other side from this direction. The hillock shape is close to a parallelogram. The rotation angle may be less (see figure 2.35). With further increasing σ the hillock orientation does not change any more, but the parallelogram vertices become sharper (figure 2.14). The crest line of the slopes passes through the hillock vertex and the parallelogram sharp vertex, which can often be seen on the crystals with an unaided eye, and the slopes become flatter. Hillocks of such shape have been observed by many scientists (Dam and van Enckevort 1984, Noor and Dam 1986).

The process of changing the hillock shapes is based on the fact that the tangential rate along the long diagonal of the parallelogram increases 5–6 times. The normal growth rate of the hillock vertex does not change. The mass of a substance crystallizing on the hillock per unit of time during its restructuring increases 2–3 times; this increase being accompanied by a decreasing steepness which is especially noticeable along the long diagonal of the rhomb where p tends to zero after a certain period of time. The rotation of the hillock is accompanied by the appearance of macrosteps on its slopes; the macrosteps are perceived as spurs on the interference fringes (see figure 2.15). After the next change in supersaturation they gradually increase, reach a maximum and disappear again.

On the ADP prismatic faces the shape of the hillocks changes also in the narrow range of σ-values. The hillocks still remain elliptical, their orientation changing only slightly, but their eccentricity is sharply increased and the steepness is diminished (see figure 2.16).

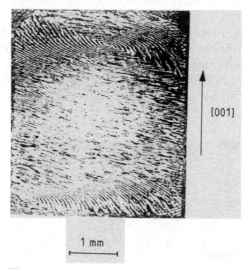

[001]

1 mm

Figure 2.14 The shape of hillocks of the KDP prismatic face at high supersaturation ($\sigma = 0.061$). In the periphery of the steep hillocks, macrosteps appear and the regularity of the interference pattern disappears.

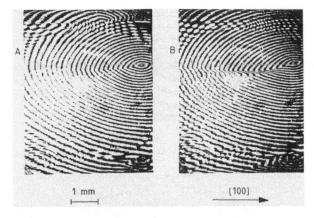

Figure 2.15 Emergence of spurs on the interference fringes with a modification of the shape of the hillocks on the KDP prismatic face with $\sigma = 0.037$. Time after establishment of supersaturation (min): A, 25; B, 45.

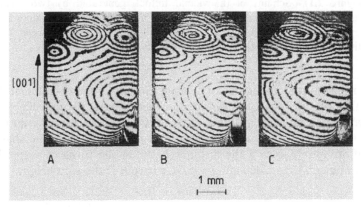

Figure 2.16 Changes of the eccentricity of hillocks on the ADP prismatic face with $t \simeq 35°C$ and $\sigma \times 10^2$: A, 4.07; B, 4.10; C, 4.22. The chromium impurity in the solution is 1.5×10^{-5} mol Cr per mol ADP.

The last section of this chapter will provide a more detailed description of the changing shapes of hillocks, some aspects of the coagulation of elementary steps into larger ones, the production of kinematic waves of steps, and the formation of macrosteps, in other words, the problem of the morphological stability of the surface is discussed.

Thus, the growth of crystals is caused by the development of dislocation hillocks on their surface. The experiments indicate that their generating lines are linear, and as a consequence, the normal (R) and tangential (v) growth rates of a hillock are connected with its steepness, p, by the geometric relation

$$R = pv \qquad (2.2)$$

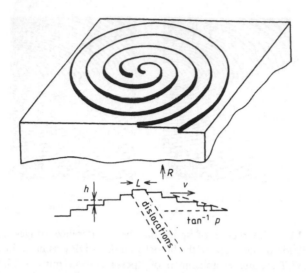

Figure 2.17 Scheme of the vicinal hillock generated by two screw dislocations with equal Burgers vectors. The lower part of the diagram represents the hillock profile where L is the distance between the dislocations and h is the elementary step height.

which is illustrated in figure 2.17. Therefore in order to understand the crystallization kinetics one should consider separately the effect of the supersaturation and that of other characteristics of the process upon p and v. The first step in this line of investigation will be to study the shape of a dislocation spiral, since if the distance between the turns of a spiral, λ, is known, one can then calculate $p = h/\lambda$, where h is the step height.

2.3   The dislocation spiral

A step on the crystal surface has a great number of kinks and grows by a normal mechanism. The velocity of its motion v is defined by the super-saturation

$$v = \beta c_0^* w\sigma = b\sigma. \tag{2.3}$$

Here w is the volume of particles in the crystal, c_0^* is the equilibrium concentration (of particles per cm^3), β is the kinetic coefficient of the step, and the term $c_0^* w$ takes into account the difference between the density of the crystal and that of the solution.

According to the Gibbs–Thomson formula, the supersaturation depends on the step curvature:

$$\Delta\mu = \Delta\mu_\infty - \alpha w/r \tag{2.4}$$

where $\Delta\mu_\infty$ is the difference of the chemical potentials of a particle in the medium and a straight-line step, α is the surface energy of a step riser, and r is the radius of step curvature. Rewriting equation (2.4) gives

$$\Delta\mu = \Delta\mu_\infty(1 - \alpha w/\Delta\mu_\infty r)$$

or, dividing both parts by $k_B T$,

$$\sigma = \sigma_\infty(1 - r_c/r). \tag{2.5}$$

Here

$$r_c = \alpha w/\Delta\mu_\infty = \alpha w/k_B T\sigma_\infty \tag{2.6}$$

is called the radius of a critical nucleus. Substituting equation (2.5) into (2.3) we obtain

$$v = v_\infty(1 - r_c/r) \tag{2.7}$$

where v_∞ is the straight-line step velocity ($r \to \infty$).

Equation (2.5) implies that the solution, which is supersaturated with respect to a straight-line step, turns out to be saturated for the step parts with a radius of curvature r_c, and for these parts $v = 0$. Generally speaking, α and β depend on the orientation of a step on the surface, and accordingly, r_c and v_∞ are anisotropic values. The importance of this fact will be discussed below.

Equation (2.7) is responsible for the curving of a step, which is formed on the surface by a screw dislocation and is originally a straight-line step, into a spiral of a certain shape. Under stationary conditions the spiral should meet the requirements that its curvature and velocity at any point (on the normal to the curve) should satisfy equation (2.7). In particular, at the point of dislocation emergence $v = 0$ and $r = r_c$. For the isotropic case (where v_∞ and r_c are constant), the characteristic features of such a spiral were studied by Burton, Cabrera and Frank (1951). We shall begin by considering a polygonal spiral since in this case it is possible to obtain an analytical solution.

2.3.1 A square spiral

Let a dislocation emerge on the surface at a point A where at the initial instant of time to step A0 is linear (figure 2.18(a)). During time τ, the step moving parallel to itself with velocity v_∞ will travel a distance $l_c = v_\infty\tau$. Here $l_c = 2r_c$ is the critical length. Upon reaching the length l_c, the step segment formed will also start moving. After a time 2τ, this moving segment forms a third segment of the same length l_c. A complete turn will be performed by the step during a time 4τ, the distance between the turns will be $\lambda = 4\tau v_\infty = 4l_c = 8r_c$. These considerations do not take into account the fact that the

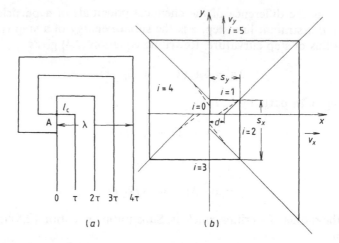

Figure 2.18   (*a*) Scheme of the formation of the square spiral; and (*b*) trajectories of its angular points. The broken lines show the initial sections of the trajectories.

step velocity depends on its length in accordance with an equation similar to (2.7):

$$v = v_\infty(1 - l_c/l). \tag{2.8}$$

Allowing for equation (2.8) leads to the fact that the first segment formed will move upwards very slowly at the beginning and the second segment of length l_c will be formed during a time greater than τ. As a result λ will be greater than $4l_c = 8r_c$. Numerical calculations carried out by Budevski *et al* (1975) indicated that, in fact, $\lambda = 19.20r_c$. There are also analytical methods to determine the shape of such a spiral and the inter-step distance (Chernov and Rashkovich 1987a,b).

One of the methods is based on calculating the trajectories of angular points of a polygonized spiral. The equation for trajectory 1 (figure 2.18(*b*)) is of the form

$$dy/dx = v_y/v_x = (1 - s_y^{-1})/(1 - s_x^{-1}). \tag{2.9}$$

Here equation (2.8) is taken into account, the dimensionless length $s = l/l_c$ ($s > 1$) is introduced, and the steps moving in the direction of corresponding axes are marked by indices x and y.

Far from the origin of the coordinates, when $s \gg 1$, v does not depend on s and, consequently, the velocities of neighbouring segments are the same. Therefore the angular points must move along the straight lines at an angle of $45°$ to the axes. These four straight lines cut off equal sections of length d on the axes; far from the origin of the coordinates the distance between neighbouring turns being $\lambda = 4d$. If v were independent of s the step would have a spiral shape as shown in figure 2.18(b) by a thick line, the value d playing the role of a critical length. The true trajectories of the angular points are represented by the curves shown in this figure by broken lines; as x, $y \to \infty$ they transform asymptotically into straight lines. The task is to determine d for the latter case. The solution of equation (2.9) will be sought in the form of subsequent approximations. The zero solution refers to v being independent of s and to linear trajectories. For the first approximation let us take the values s_x and s_y that follow from the trivial case, in other words, from the position of the spiral angles on the curves given in figure 2.18(b). The values s_x and s_y, corresponding to the trajectories found from the first trajectories, will become the second approximation, and so on. In practice, the first approximation turns out to be sufficient.

For a face with square symmetry all four trajectories are identical. In solving equation (2.9) it should be taken into account that s_x and s_y are expressed differently by x and d for different segments. Indeed

with $\qquad 1 \leqslant x \leqslant d \qquad s_y = x \qquad\qquad s_x = x + d$

with $\qquad d \leqslant x \leqslant 2d \qquad s_y = x \qquad\qquad s_x = 2x \qquad\qquad$ (2.10)

with $\qquad x \geqslant 2d \qquad s_y = 2x - 2d \qquad s_x = 2x.$

By solving equation (2.9) with allowance for (2.10), and by passing values into the trajectory equation obtained in the limit $x \to \infty$ whilst requiring that this limiting trajectory should coincide with the curve $y = x - d$, we obtain

$$d = 1 - \frac{1}{2} \ln\left(\frac{2d-1}{4d-1}\right) + \frac{d}{2d-1} \ln\left(\frac{4d-1}{2d}\right) + \frac{d}{d-1} \ln\left(\frac{d^2}{2d-1}\right). \quad (2.11)$$

The root of equation (2.11) is $d = 2.52262$, i.e. $\lambda = 4dl_c = 10.090l_c = 20.18r_c$, which differs only slightly (by $\simeq 5\%$) from the exact value of $\lambda = 19.20r_c$.

Let us consider the second method of calculating the distance between the spiral turns. Designating the segments as shown in figure 2.18(b) and introduce a dimensionless velocity

$$u(s_i) = u_i = v(s_i)/v_\infty = 1 - s_i^{-1} \quad (2.12)$$

with time t measured in units of l_c/v_∞. With the aid of figure 2.18(b) one can easily write for the velocities of segments of different lengths of the spiral:

$$ds_0/dt = u_1 \qquad ds_1/dt = u_2 \qquad ds_i/dt = u_{i-1} + u_{i+1} \qquad i = 2, 3, \ldots. \quad (2.13)$$

Now, the task is to determine the time τ taken by the spiral to rotate by $90°$, i.e. for its zero segment ($i = 0$) to attain a critical length l_c, i.e. $s_0(\tau) = 1 = s_1(0)$. Similarly, after time τ, segment i becomes the $(i+1)$th

$$s_1(\tau) = s_2(0), \ldots, s_i(\tau) = s_{i+1}(0). \quad (2.14)$$

If τ is known, then the time taken by the spiral to rotate equals 4τ and the distance between its turns far from the centre is $\lambda = 4\tau l_c$.

Integrating the system of contacting equations (2.13) going from the segments with $i \gg 1$ to the spiral centre can be easily done, if one employs the approximation $u_{i-1} + u_{i+1} = 2u_i$ for $i \geq 2$. Then the last equation in (2.13) can also be integrated with $t = \tau$ and yields, for $i = 2, 3, \ldots$, the following:

$$F(s_{i\tau}) - F(s_{i+1,\tau}) = 2\tau \qquad F(s) \equiv s + \ln(s-1) \qquad s_{i\tau} \equiv s_i(\tau). \quad (2.15)$$

In this approximation $ds_1/ds_2 = u_2/(u_1 + u_3) \simeq 1/2$, i.e.

$$s_2(\tau) - s_2(0) = 2(s_1(\tau) - s_1(0)) \qquad s_2(\tau) - s_2(t) = 2(s_1(\tau) - s_1(t))$$

or, allowing for equation (2.14),

$$s_{2\tau} = 3s_{1\tau} - 2 \qquad s_2 = 2s_1 + s_{1\tau} - 2. \qquad (2.16)$$

The condition for the zero segment to attain a critical length $s_0(\tau) = 1$ will be obtained if the first two equations of (2.13) are taken into account, and hence it follows that $ds_0 = (u_1/u_2)ds_1$:

$$1 = \int_1^{s_{1\tau}} (u_1/u_2)ds_1 = \int_1^{s_{1\tau}} (1 - s_1^{-1})/(1 - s_2^{-1})ds_1. \qquad (2.17)$$

The integral in equation (2.17) allowing for (2.16) provides the following equation for $s_{1\tau}$:

$$s_{1\tau} - 2 - [(s_{1\tau} - 2)/(s_{1\tau} - 3)]\ln(s_{1\tau}) + [(s_{1\tau} - 1)/2(s_{1\tau} - 3)]\ln 3 = 0. \quad (2.18)$$

Its root is $s_{1\tau} = 2.836$. But, substituting $s_{2\tau}$ from equation (2.16) into (2.15) we obtain

$$\tau = s_{1\tau} - 1 + 0.5 \ln 3 = 2.385.$$

Subsequently $\lambda = 4\tau l_c = 9.54 l_c = 19.08 r_c$, which is close to the exact value of $\lambda = 19.20 r_c$.

Budevski *et al* (1975) calculated λ for triangular and hexagonal spirals and obtained $\lambda = 18.85 r_c$ and $19.02 r_c$, respectively. These results can be obtained using the methods described above. Since the approximate, analytical, solution is close to the exact, numerical, one, it indicates that it is only on the first, or at most on the second turn of a spiral that the Gibbs–Thomson effect is manifested noticeably, while on the remainder, the main part of the spiral λ is constant. Therefore, at a distance of $\simeq 19 r_c$ from the hillock vertex, the slope steepness must be constant over its entire periphery and depends linearly on the supersaturation:

$$p = h/\lambda = h/19 r_c = (h k_B T/19 \alpha w)\sigma. \qquad (2.19)$$

2.3.2 The isotropic spiral

In this case, when the stationary form conforming to equation (2.7) has been established, the spiral turns as a whole around its centre and its curvature at a given distance from the centre is independent of time.

Two methods of calculating the spiral shape are known. The first one was realized in the first approximation by Burton *et al* (1951), and later the exact solution was obtained by Cabrera and Levine (1956) and by Müller–Krumbhaar *et al* (1977). The second method was suggested by Mikhailov *et al* (1989).

In the polar system of coordinates (ρ, θ) the spiral stationary equation $\theta(\rho)$ is

$$\theta = f(\rho) + \omega t$$

where ω is the spiral rotation angular velocity (to be exact, the clockwise rotation), and t is the time. If ω is known, then the normal growth rate R of a hillock and the distance between the turns λ at a sufficient distance from the centre are

$$R = \omega h/2\pi \qquad \lambda = 2\pi v_\infty/\omega. \tag{2.20}$$

In polar coordinates, the velocity v of a spiral moving along the normal and the curvature radius r are given by the formulae

$$v = \frac{\omega\rho}{[1 + (\rho\dot\theta)^2]^{1/2}} \qquad r = \frac{[1 + (\rho\dot\theta)^2]^{3/2}}{2\dot\theta + \rho\ddot\theta + \rho^2\dot\theta^3}$$

where $\dot\theta = \partial\theta/\partial\rho$ and $\ddot\theta = \partial^2\theta/\partial\rho^2$. Substituting v and r into equation (2.7) we obtain a non-linear differential equation for the determination of $\theta(\rho)$. Its approximate solution can be obtained with $\rho \to \infty$ and $\rho \to 0$. Let us introduce the dimensionless values s and ω_1:

$$\rho = sr_c \qquad \omega = \omega_1 v_\infty/r_c. \tag{2.21}$$

For large s, $v = \omega_1 v_\infty/\theta'$ ($\theta' = \partial\theta/\partial s$), the curvature radius being equal to the radius vector: $r = sr_c$. Substituting this into equation (2.7) we obtain for $s \to \infty$

$$\theta' = \omega_1(1 - s^{-1}). \tag{2.22}$$

For small s, $v = \omega_1 v_\infty s_1$, and $r_c/r = 2\theta' - \omega_1 s/3$. Substituting this into equation (2.7) gives for $s \to 0$

$$\theta' = 0.5 - \omega_1 s/3. \tag{2.23}$$

Choosing a solution of the form

$$\theta' = a + b/(1 + cs) \tag{2.24}$$

and defining the constants in equation (2.24) and ω_1 by imposing a requirement that equation (2.24) should transform into (2.22) and (2.23) for large and small s. Then

$$a = b/c = \omega_1 \qquad a + b = 1/2 \qquad bc = \omega_1/3.$$

Hence,

$$\omega_1 = a = \sqrt{3}/2(1 + \sqrt{3}) = 0.315 \qquad b = 1/2(1 + \sqrt{3}) \qquad c = 1/\sqrt{3}$$

$$\theta' = \omega_1[1 + (\sqrt{3} + s)^{-1}] \qquad \theta = \omega_1[s + \ln(1 + s/\sqrt{3})].$$

This solution obtained by Burton *et al* (1951) satisfies equation (2.7) to an accuracy of a few per cent. The exact value of $\omega_1 = 0.331$ was obtained by numerical calculation (Müller–Krumbhaar *et al* 1977). Allowing for equations (2.20) and (2.21)

$$\lambda = 2\pi r_c/\omega_1 = 18.98 r_c.$$

Note that Cabrera and Levine (1956) reduced the order of the differential equation for a spiral. They took into account that $s\theta' = \tan \psi$, where ψ is the angle between the radius vector and the tangent to a spiral (see figure 2.19). Then

$$v = \omega_1 v_\infty s/\sec \psi \qquad r_c/r = (d\psi/ds + s^{-1} \tan \psi)/\sec \psi.$$

Substitution into (2.7) yields the solution desired

$$\psi' = \sec \psi - s^{-1} \tan \psi - \omega_1 s. \tag{2.25}$$

Mikhailov *et al* (1989) obtained the isotrope spiral equation in a more general form, which admits curvature dependences of v other than those of equation (2.7).

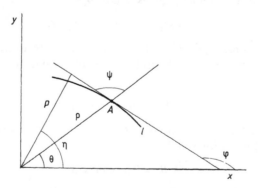

Figure 2.19 A section of the spiral in different coordinate systems. ρ, θ—the radius-vector of point A and the polar angle; ψ—the angle between the curve tangent and the radius-vector; φ—the angle between the tangent and the x axis; p, η—the distance from the origin of the coordinates to the tangent and the corresponding angle.

Let us define an equation for the step shape evolution with time t by assuming this shape to be the dependence $k = k(l, t)$ of the curvature $k = r^{-1}$ on a length l along the curve.

Let us consider a small part of a step and draw normals to it at every point. After the time interval dt, each point would move along its normal by a distance $v\,dt$ converting the initial part of the step into a new one. Let us choose a polar system of coordinates (ρ, θ) in such a way that its origin should be at the centre of curvature for one of the points a of the initial part: $\rho_a = r_a = k_a^{-1}$. After the time interval dt, point a would drift to point b, and hence

$$\rho_b = \rho_a + v\,dt. \tag{2.26}$$

The steepness k_b of a new part at point b can be calculated by substituting equation (2.26) into the equation for the curve curvature

$$k = (\rho^2 + 2\rho'^2 - \rho'')/(\rho^2 + \rho'^2)^{3/2}.$$

Conserving the first-order terms over dt and taking into account that $\rho_a' = d\rho_a/d\theta = 0 = \rho_a''$ then

$$k_b = k_a - vk_a^2\,dt - k_a(d^2v/d\theta^2)\,dt.$$

Introducing the arc length l so that $dl = \rho\,d\theta$ we obtain

$$dk = k_b - k_a = -(k_a^2 v + d^2v/dl^2)\,dt. \tag{2.27}$$

On the other hand, the increment in the curvature is

$$dk = (\partial k/\partial l)\,dl + (\partial k/\partial t)\,dt. \tag{2.28}$$

For the arc length increment during a time dt with a normal drift of each part with velocity v we obtain

$$dl = \int_0^\theta (v\,dt)\,d\theta = \int_0^l k(v\,dt)\,dl = \left(\int_0^l kv\,dl\right)dt. \tag{2.29}$$

In fact, the curve is divided into a set of small parts and for each part a polar system of coordinates of its own was introduced. The total increment in the curve arc length is given by the sum of the increments at separate parts.

Substituting equations (2.27) and (2.29) into (2.28) we have

$$\partial k/\partial t + \left(\int_0^l kv\,dl\right)\partial k/\partial l = -k^2v - \partial^2v/\partial l^2. \tag{2.30}$$

Assuming the normal drift-velocity of a step to be uniquely defined by its local curvature at the corresponding point we have $v = v(k(l, t))$.

Equation (2.30) completely defines the step evolution as well as the crystallization front perturbation in a thin flat layer. A rigorous deduction of (2.30) was made by Brazhnik *et al* (1988).

Equation (2.30) should be supplemented by boundary and initial conditions. Since the step start ($l = 0$) is fixed by a dislocation, then the curvature at that point should equal the critical one $k(0) = k_c = r_c^{-1}$. At large distances from the centre with $l \to \infty$, $k \to 0$.

In order to study the stationary case when a spiral does not change its shape but turns with time, we shall employ equation (2.30). If this is the case, then v and k_c are independent of the orientation on the surface, which was implied in deriving equation (2.30). In this case $k = k(l)$ and (2.30) reduces to

$$(\mathrm{d}k/\mathrm{d}l) \int_0^l kv \, \mathrm{d}l = -k^2 v - \mathrm{d}^2 v/\mathrm{d}l^2.$$

This equation can easily be integrated once more thus taking the form

$$k \int_0^l kv \, \mathrm{d}l + \mathrm{d}v/\mathrm{d}l = \omega. \qquad (2.31)$$

Differentiating equation (2.31) proves the validity of this integration. The integration constant ω is equal to the spiral rotation angular velocity. This follows from the fact that $\mathrm{d}v/\mathrm{d}l$, with $l = 0$, is the angular velocity.

Since $v = v(k)$, equation (2.31) is a first-order integro-differential equation relative to $k = k(l)$. Its solution must satisfy two boundary conditions:

$$k(0) = k_c \qquad \text{and} \qquad \lim_{l \to \infty} k(l) = 0.$$

These conditions can only be satisfied with a particular value of the constant ω. In this way the spiral rotation angular velocity is defined.

An important feature of equation (2.31), which distinguishes it from (2.25), is the fact that (2.31) does not imply a certain type of $v(k)$.

Let us consider a case where equation (2.7) is valid:

$$v = v_\infty(1 - k/k_c). \qquad (2.7a)$$

Now, introducing dimensionless values of the type of equation (2.21) gives

$$y = k/k_c \qquad x = lk_c \qquad \omega_1 = \omega/v_\infty k_c \qquad (2.32)$$

and then substituting (2.7a) into (2.31) we have

$$y \int_0^x y(1 - y) \, \mathrm{d}x - \mathrm{d}y/\mathrm{d}x = \omega_1 \qquad (2.33)$$

with boundary conditions $y(0) = 1$ and $y(\infty) = 0$.

Numerical solution of equation (2.33) yields $\omega_1 = 0.3310833$.

Solutions of equation (2.31) for other dependences $v(k)$ will be dealt with when impurity effects are discussed in section 2.5.

2.3.3 Account of anisotropy

The step riser surface energy α and the kinetic coefficient β both depend on the step orientation on the surface φ (see figure 2.19); in accordance with equations (2.6) and (2.3), r_c and v_∞ depend on φ as well.

Given the shape of a critical nucleus, $\alpha(\varphi)$ can be defined by the following equation (Herring 1951)

$$\alpha + \mathrm{d}^2\alpha/\mathrm{d}\varphi^2 = \Delta\mu r/\omega$$

where $r = r(\varphi)$ is the critical nucleus curvature radius. Unfortunately, the technique used in our experiments does not allow us to determine the dislocation hillock shape near its centre (at a distance of $\simeq r_c$); other techniques available also do not permit the anisotropy of α to be ascertained.

This is not the case with the anisotropy of β. It is defined by the crystal growth shape (Chernov 1971):

$$v(\varphi) + \mathrm{d}^2v(\varphi)/\mathrm{d}\varphi^2 = vr(\varphi) \tag{2.34}$$

where $r(\varphi)$ is the growth shape curvature radius and v is the constant inversely proportional to time. Since the hillock shape far from the centre is known, then $v_\infty(\varphi)$ (or $\beta(\varphi)$) can easily be calculated. For example, for an ellipse when eccentricity $e = (1 - a_y^2/a_x^2)^{1/2}$, where a_x and a_y are large and small half-axes, the curvature radius as a function of φ is

$$r(\varphi) = (a_y^2/a_x)(1 - e^2\cos^2\varphi)^{-3/2}. \tag{2.35}$$

Substituting equation (2.35) into (2.34) gives $v(\varphi)$ in the form

$$v(\varphi) = C_1(\varphi)\cos\varphi + C_2(\varphi)\sin\varphi.$$

The unknown coefficients C_1 and C_2 can readily be found by the variation method. The solution of equation (2.34) for this case has the form

$$v(\varphi) = va_x(1 - e^2\cos^2\varphi)^{1/2}. \tag{2.36}$$

To study the kinematics of a growth spiral allowing for the anisotropy of v_∞ and r_c is a complicated mathematical problem. With respect to solution simplicity, elliptical anisotropy turns out to be an exception, assuming that the shapes of growth and equilibrium are equal, i.e. anisotropies α and β coincide (Avanesyan 1988).

Let us consider a step in the coordinate system (p, η) shown in figure 2.19. Note that earlier this system of coordinates was employed by Burton *et al* (1951) to prove the Wulff theorem. In coordinates of (p, η) the spiral evolution equation has a particularly simple form. Indeed,

$$v = \partial p/\partial t \qquad k = (p + \partial^2 p/\partial\eta^2)^{-1}. \tag{2.37}$$

Substituting equation (2.37) into (2.7a) we have

$$\partial p/\partial t = v_\infty[1 - k_c^{-1}(p + \partial^2 p/\partial\eta^2)^{-1}] \tag{2.38}$$

where v_∞ and k_c are functions of η.

In the isotropic case, k_c and v_∞ are constant. Proceeding to the dimensionless variables $t' = k_c v_\infty t$ and $p' = k_c p$ and omitting the prime, we shall come to a non-linear equation in partial derivatives for the function $p(\eta, t)$:

$$\partial p/\partial t = 1 - (p + \partial^2 p/\partial \eta^2)^{-1}. \tag{2.39}$$

The stationary solution of equation (2.39) is sought under the following boundary conditions: $p = 0$ with a certain value of η that is independent of t, and $p \to \infty$ with $\eta \to \infty$.

We shall show that equation (2.39) can be reduced to (2.25). Under stationary conditions with constant k_c and v_∞, $p(\eta, t)$ transforms into $p(\eta + \omega_1 t)$ and equation (2.39) can be rewritten as

$$\omega_1 \, dp/d\eta = 1 - (p + \partial^2 p/\partial \eta^2)^{-1}.$$

It follows from figure 2.19 that $\rho^2 = p^2 + (dp/d\eta)^2$ (here ρ and p are measured in units of r_c), hence $dp/d\eta = (\rho^2 - p^2)^{1/2}$ and

$$(\partial p/\partial \eta)(p^2 + d^2 p/d\eta^2) = \rho \, d\rho/d\eta \qquad \text{or} \qquad (p + d^2 p/d\eta^2)^{-1} = dp/\rho \, d\eta.$$

Consequently

$$\omega_1 (\rho^2 - p^2)^{1/2} = 1 - dp/\rho \, d\eta.$$

Since $p = \rho \sin \psi$, $dp/d\rho = \sin \psi - \rho \cos \psi \, d\psi/d\rho$, we come to equation (2.25)

$$d\psi/d\rho = \sec \psi - \rho^{-1} \tan \psi - \omega_1 \rho.$$

For the case of elliptic anisotropy, $k_c(\varphi)$ and $v_\infty(\varphi)$ are given by equations (2.35) and (2.36). Let a_{cx} and a_{cy} be the critical nucleus ellipse half-axes. Since $\eta = \pi/2 - \varphi$ we obtain

$$\begin{aligned} v_\infty(\eta) &= va_{cx}(1 - e^2 \sin^2 \eta)^{1/2} \\ k_c(\eta) &= (a_{cx}/a_{cy}^2)(1 - e^2 \sin^2 \eta)^{3/2} \end{aligned} \tag{2.40}$$

Returning to equation (2.38) and changing from p to $\hat{p} = vp/v_\infty$ and from t to $\hat{t} = vt$, then instead of (2.38) we have

$$d\hat{p}/d\hat{t} = 1 - \left(\hat{p} \frac{k_c}{v}(v_\infty + d^2 v_\infty/d\eta^2) + 2\frac{k_c}{v}(dv_\infty/d\eta)(\partial \hat{p}/\partial \eta) + \frac{k_c}{v} v_\infty(\partial^2 \hat{p}/\partial \eta^2) \right).$$

In view of both equation (2.34) and the assumption that when growing a critical nucleus does not change its shape, the first summand between large brackets above is equal to $\hat{p}$. Now instead of η we shall introduce a variable.

$$\hat{\eta} = v^{1/2} \int_0^\eta (v_\infty k_c)^{-1/2} \, d\eta = \int_0^\eta [a_{cx}(1 - e^2 \sin^2 \eta)/a_{cy}]^{-1} \, d\eta$$

$$= \tan^{-1}[(1 - e^2)^{1/2} \tan \eta].$$

Then the second summand in large brackets will take the form $(\mathrm{d}\ln v_\infty^2/\mathrm{d}\hat{\eta})(\partial\hat{p}/\partial\hat{\eta})$, and the third summand will be equal to

$$\partial^2\hat{p}/\partial\hat{\eta}^2 - [\mathrm{d}\ln(k_c v_\infty)^{1/2}/\mathrm{d}\hat{\eta}](\partial\hat{p}/\partial\hat{\eta}).$$

Merging the terms with $\partial\hat{p}/\partial\hat{\eta}$ we shall see that the coefficient of this derivative is equal to $\mathrm{d}\ln(v_\infty^3/k_c)^{1/2}/\mathrm{d}\hat{\eta}$ and is therefore equal to zero, since in view of equation (2.40) $v_\infty^3/k_c = v^3 a_{cx}^2 a_{cy}^2 = $ constant. Thus

$$\partial\hat{p}/\mathrm{d}\hat{t} = 1 - (\hat{p} + \partial^2\hat{p}/\partial\hat{\eta}^2)^{-1}$$

which coincides with equation (2.39).

Thus, the transformations

$$\hat{p} = p/a_{cx}(1 - e^2\sin^2\eta)^{1/2} \qquad \hat{\eta} = \tan^{-1}[(1 - e^2)^{1/2}\tan\eta] \qquad \hat{t} = vt \quad (2.41)$$

lead the spiral equation (2.38) to an isotropic form (2.39). Therefore the distance between the spiral turns along the axes x and y at a sufficient distance from the centre will be equal to

$$\lambda_x = 2\pi a_{cx}/\omega_1 \qquad \lambda_y = 2\pi a_{cy}/\omega_1$$

ω_1 having the same volume as in the isotropic case. For steps oriented at an angle φ (propagating at an angle η)

$$\lambda(\varphi) = 2\pi r_c/\omega_1 \qquad r_c = a_{cy}^2/a_{cx}(1 - e^2\cos^2\varphi)^{3/2} = (\alpha + \mathrm{d}^2\alpha/\mathrm{d}\varphi^2)\omega/k_B T\sigma.$$
$$(2.42)$$

Note that transformations (2.41) are none other than an expansion transformation $(1 - e^2)^{-1/2} = a_{cy}/a_{cx}$ times along the y axis and a simple alteration of the scale.

Chernov *et al* (1987) considered the surface energy anisotropy effect and the kinetic coefficient for a rectangular spiral. This can be done using both the trajectory method and the approximate analytical technique described above. The results are represented in figure 2.20. Relation $B = v_{\infty y}/v_{\infty x}$ is plotted on the abscissa, λ_x/λ and λ_y/λ are plotted on the ordinate, where λ is the interstep distance for a square spiral ($B = 1$). The curves numbered 1 (marked by indices x and y for λ_x and λ_y, respectively) refer to the isotropic surface energy of steps, implying that $\alpha_x = \alpha_y$. The curves numbered 2 describe the growth hillock anisotropy, when the kinetic coefficient is proportional to the surface energy. As in the case of an elliptic spiral, this assumption results in the same values of λ as for the isotropic case. The curves numbered 3 characterize the opposite relation: the greater the step kinetic coefficient, the lower the step surface energy. The latter relation is based on the assumption that the free surface energy of a step riser decreases as the density of kinks on it increases and, consequently, the kinetic coefficient becomes greater.

From figure 2.20 it follows that the anisotropy of the shape of the main macroscopically extended part of a hillock, where the Gibbs–Thomson effect

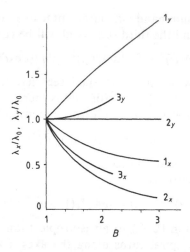

Figure 2.20   The anisotropy effect of the kinetic coefficient $B = v_{\infty y}/v_{\infty x}$ on the interstep distance along the axes x and y with a different anisotropy of the surface energy of the step face: l_{cy}/l_{cx} equals 1 (curves numbered 1), B (curves numbered 2), B^{-1} (curves numbered 3).

is small, is defined by the kinetics of steps only. At the same time the absolute steepness is proportional to α; however, it is impossible to determine the surface energy anisotropy by the shape of a hillock at a great distance from its vertex.

It should also be noted that, if $B = 1$ and l_{cy}/l_{cx} is plotted on the abscissa in figure 2.20, then the curve 1_y would correspond to the ratio $\lambda_x = \lambda_y$ to λ is the isotropic case. Thus, even if a hillock is a circular cone, the anisotropy α is revealed appreciably in its steepness.

To summarize, the considerations presented in this section show that dislocation hillocks generated by single dislocations should have linear generating lines whose slope to a singular face is proportional to σ. In the isotropic case and in the case where the kinetic coefficient is proportional to the free surface energy of the step riser, the interstep distance is $\lambda \simeq 19r_c$. This is valid not only for circular and rectangular spirals but for triangular and hexagonal ones as well (Budevski *et al* 1975). The hillock shape is defined by the kinetic coefficient anisotropy and the latter can be determined by experiment.

2.4   The main characteristics of dislocation growth

In this section we shall discuss the parameters characterizing the crystal growth rate as a function of the supersaturation in the absence of impurities (their effect is dealt with in section 2.5). These parameters are the dislocation source structure and its activity, the free surface energy and the kinetic coefficient.

2.4.1 Dislocation source activity

Burton *et al* (1951) thoroughly studied the topology of a surface with many dislocations capable of generating growth hillocks. They showed that if the distance between dislocations is greater than half of the distance between the steps generated by one dislocation (i.e. greater that $9.5r_c$), then each dislocation produces a hillock of its own. Otherwise, when the distance between dislocations (in a dislocation pair or group) is less than $9.5r_c$, the spirals unite and become multiple. If a group of m screw dislocations of the same sign lie on a line of length L (this arrangement is quite typical, since dislocation groups often arise at the boundaries of blocks), then the spiral turn-time is composed of the time taken by one dislocation to turn around the centre $19r_c/v_\infty$ and the time required for passing around section L in both directions ($2L/v_\infty$). In this case each turn would make the hillock height m times greater than the turn produced by a single spiral. Thus, in this case

$$\lambda = (19r_c + 2L)/m \tag{2.43}$$

$$p = mh/(19r_c + 2L). \tag{2.44}$$

If the steps are of different height, then mh is the summary Burgers vector of a dislocation source. When dislocations are not arranged on a straight line, $2L$ stands for the perimeter of an area occupied by a dislocation group under study.

When deriving equation (2.43) the dependence $v(r)$ was not taken into account when the step passed around the dislocation source. When the case of a square spiral passing around a linear obstacle was considered using the trajectory method, it was shown that the true value of λ exceeds the one that follows from equation (2.43) by a mere 10% for the supersaturation case at which $19r_c = 2L$, and the error should be less at other supersaturations. Hence, if there are two dislocations in the source, the error is a maximum at $19r_c = 2L$, and diminishes with smaller values of σ, and with a greater value of σ the source breaks into two. With $m > 2$ the error in using the calculations of equation (2.43) is always less than 10%.

The ratio of the normal growth rate $R = pv$ of the generated hillock to the value of R of the hillock generated by a single dislocation with a singular Burgers vector is called the dislocation growth source activity ε. Assuming that v is independent of p, we conclude that ε is the ratio of equation (2.44) to (2.19)

$$\varepsilon = m/(1 + 2L/19r_c). \tag{2.45}$$

It follows from equation (2.45) that, with $\sigma \ll 1$, ε depends linearly on the supersaturation and decreases as the latter increases. The source activity with increasing σ may be less than unity if $m - 1 < 2L/19r_c$ (given the condition that $m > 2L/19r_c$, otherwise the source decomposes).

The principal peculiarity of equation (2.44), which characterizes the hillock steepness as a function of the supersaturation, is that the linearity of $p(\sigma)$ is only revealed for small σ when $19r_c \gg 2L$. For large values of σ, when $19r_c \ll 2L$, the dependence of p on σ vanishes. It should be pointed out, however, that for $19r_c \leqslant 2L/m$ the source breaks into several individual ones.

Dislocation hillocks generated on single dislocations are rare. This is due to the fact that solution inclusions resulting from the nucleus regeneration are the principal dislocation source. As a rule, bunches of dislocations are produced on these inclusions. A hillock generated on a single dislocation is shown in figure 2.9. Figure 2.21 gives the dependences of p, v and R on σ for this hillock. One can see that p depends linearly on σ. Further x-ray topographic investigation showed that a single dislocation was the growth centre.

Figure 2.22 represents a photograph of a face with three hillocks. For two of them (marked with a cross and a circle) the dependences of p, v and R on σ were measured and are shown in figure 2.23.

Just as in figure 2.21, v depends linearly on σ, the experimental points for both hillocks lying on the same straight line. The dependence $p(\sigma)$ is different for each hillock, but both curves intersect and they are not straight lines in contrast to those in figure 2.21. Rewriting equation (2.44) taking into account the expression for r_c from equation (2.6) gives

$$p^{-1} = 2L/mh + (19\alpha w/mhk_B T)\sigma^{-1}. \qquad (2.46)$$

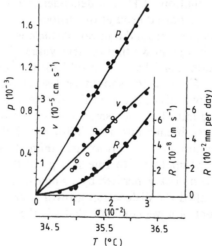

Figure 2.21 Effect of σ on the vicinal slope p, the tangential v and the normal growth rate R of the hillock generated by one dislocation (as in figure 2.9).

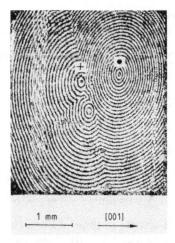

Figure 2.22 Three dislocation sources on the ADP prismatic face.

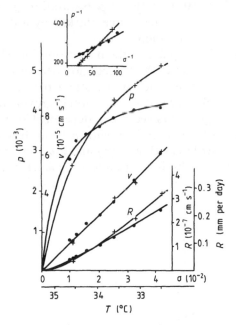

Figure 2.23 The effect of the supersaturation on the characteristics of two hillocks generated by a group of dislocations. ● and + refer to the hillocks shown in figure 2.22.

If equation (2.44) is correct, then in coordinates (p^{-1}, σ^{-1}) the experimental points should lie on straight lines. These lines are plotted in the inset of figure 2.23. They help to determine the characteristics of a dislocation source and its activity. To do this one should know the value of $19\alpha w / h k_{\mathrm{B}} T$ that is equal to the cotangent of the slope angle of the straight line $p(\sigma)$ for a hillock generated on a single dislocation. From figure 2.21 we have

$$p = h/19r_c = (6.3 \pm 0.2) \times 10^{-2}\sigma. \tag{2.47}$$

Processing the data of figure 2.23 gives

$$p_\bullet^{-1} = (208 \pm 5) + (1.346 \pm 0.009)\sigma^{-1} \pm 5.8$$
$$p_+^{-1} = (134 \pm 3) + (2.587 \pm 0.005)\sigma^{-1} \pm 2.7.$$

Making use of equations (2.46) and (2.47) gives

$$L_\bullet = 0.22 \pm 0.06\,\mu m \qquad m_\bullet = 11.8 \pm 1.2$$
$$L_+ = 0.31 \pm 0.04\,\mu m \qquad m_+ = 6.1 \pm 0.2.$$

The dependence of the activity of these sources on σ is given in figure 2.24. For the ADP prismatic face $h = 7.5 \times 10^{-8}$ cm, and making use of equation (2.47) we have $19r_c = 1.2 \times 10^{-6}/\sigma$, therefore the source marked by a cross breaks down with a value of the supersaturation greater than $19r_c = 1.2 \times 10^{-6}\,m/2L \simeq 0.12$, and the source marked by a dot will persist even with much higher supersaturations.

Different changes in steepness of the dislocation hillocks associated with their different activities causes the competition demonstrated in figure 2.10. Indeed, it is clear from geometric considerations that in the competition for the surface the hillock survives whose normal growth rate is greater, i.e. whose value of p is greater (since v is independent of p, the proof of the statement is given below). Therefore, with σ being below the intersection point of the curves $p(\sigma)$ (as well as $R(\sigma)$ and $\varepsilon(\sigma)$) in figures 2.23 and 2.24 the hillock marked by a dot is stable, and the hillock marked by a cross is

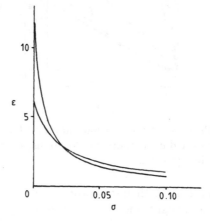

Figure 2.24   The effect of σ on the activity of the dislocation sources generating the hillocks shown in figure 2.22.

stable under greater supersaturation. From these considerations it follows that the process is reversible.

The alteration of the leading growth centres with varying σ was observed by many scientists; Kasatkin (1964) was one of the first to observe it. Bozin and Smecerovic (1983) made an attempt to determine $\varepsilon(\sigma)$ for the ADP prismatic faces for small crystals (0.02–0.2 mm) which are produced during spontaneous crystallization. The results obtained were not conclusive owing to imperfections in the techniques used and the inadequacy of the underlying principles.

Two inequalities that must be satisfied simultaneously are the condition for the intersection of $p(\sigma)$ of the two hillocks: if $m_2 > m_1$, then $m_2/L_2 < m_1/L_1$. If the second inequality is not satisfied, then over the whole range of supersaturations the source, whose value of m is greater, is stable. However, even in this case, with increasing σ the hillocks that are not stable under stationary conditions may appear temporarily on the face. With a further decrease in σ they do not appear. Figure 2.25 illustrates the mechanism of this process.

Suppose that with a given value σ' there is a hillock of steepness p_1' on the surface. Let the second hillock, whose vertex is at a distance l from that of the first one, have a steepness $p_2' < p_1'$ under the same supersaturation and as a result it cannot exist on the surface. If p_1 is less than the maximum possible steepness of the second hillock, then as the supersaturation increases up to σ'' (with the corresponding tangential velocity v'') the steepness of the hillocks increases up to p_1'' and p_2'', and if $p_2'' > p_1'$, then during a certain time τ the second hillock appears on the slope of the first one when the slope is still steep. The time

$$\tau = (l/v'')[1 + (p_2'' - p_1')/(p_1'' - p_2'')]$$

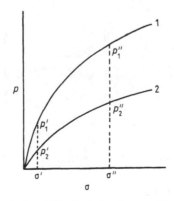

Figure 2.25 Plot of the hillock behaviour with increasing supersaturation.

can be much greater than the time that would be taken by the slope of the first hillock to move a distance l in the absence of the second source (i.e. it is possible that $(p_2'' - p_1') > (p_1'' - p_2'')$). The latter circumstance is caused by the fact that the intersection line of the slopes moves with a velocity v that is less than v

$$v = v(p' - p'')/(p' + p'').$$

It has already been mentioned that with changing σ the source structure may also be changed: its activity may be either increased or decreased. Even with constant σ the source may break down, since as the crystal grows the distance between separate dislocations in their diverging bunch becomes greater. One of the possible situations is shown in figure 2.26. One can see that as σ increases a new steep hillock appears (frames A–C), but with a further increase in σ its steepness decreases (frames D–G). On decreasing σ, that hillock does not reappear (frame H). The structure of the source generating that hillock has evidently been altered.

2.4.2   Free surface energy

The free surface energy α (together with the kinetic coefficient β) is a fundamental characteristic of the crystal–solution system. The value of α determines the size of a critical nucleus, the steepness of a dislocation hillock, the activity of a dislocation source and, ultimately (together with β) the crystal growth rate.

To determine the surface energy of the interface for solid bodies is a difficult task. In this connection experimental dependences $p(\sigma)$ offer unique

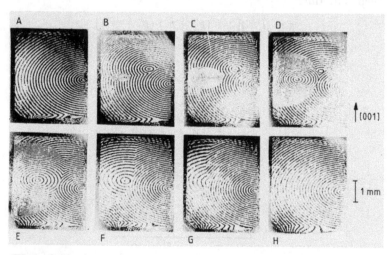

Figure 2.26 Anomalous decrease of the dislocation hillock slope with increasing supersaturation for the ADP prismatic face. $\sigma \times 10^2$: A, 1.26; B and C, 1.35; D, 1.46; E, 1.58; F, 1.71; G, 1.82; H, 1.48.

opportunities. The problems arising due to the anisotropy α have been discussed in the previous section. Taking these above considerations into account and assuming for elliptical hillocks that $\alpha \sim \beta$, it is then possible to determine α for prismatic faces of ADP and KDP crystals if the hillocks are produced by single dislocations with a known Burgers vector.

For ADP, equation (2.47) gives $p(\sigma)$ along $[100]$ as

$$p/\sigma = hk_B T/19\alpha_x w = (6.3 \pm 0.2) \times 10^{-2}.$$

Using the most probable singular Burgers vector $h = 7.53$ Å, a molecular volume $w = 10.60 \times 10^{-23}$ cm^3 and $T = 306.5$ K we find for the steps that are parallel to $[001]$ $\alpha_z = 25.1$ erg cm^{-2}. Since the ratio of the hillock slopes along $[100]$ and $[001]$ is $B = 1.60 \pm 0.07$, then for the steps that are parallel to $[100]$ we have $\alpha_x = (16.1 \pm 0.5)$ erg cm^{-2}.

The curves in figure 2.20 allow us to estimate the error in the calculation of α, if the assumption of α and β being proportional is wrong. One can see that if α is inversely proportional to β, then the indices of α_x and α_y should be replaced and the numerical values would be correct to an accuracy of $\simeq 5\%$. But if α is isotropic, then $\alpha_x = \alpha_y = 20$ erg cm^{-2} (the minimum value of α being 25% greater, the maximum one being 25% less).

The dependence $p(\sigma)$ for a hillock on a singular dislocation on the KDP prismatic face is given in figure 2.28. For the initial, linear section of the curve $p/\sigma = 0.0945 \pm 0.0083$. The ratio of the slopes along $[100]$ and $[001]$ is $B \simeq 1.35$, $w = 9.68 \times 10^{-23}$ cm^3 and $T = 301$ K. Hence,

$$\alpha_x = (16.7 \pm 1.5) \, \text{erg cm}^{-2} \qquad \alpha_y = 22.5 \, \text{erg cm}^{-2}.$$

Assuming $\alpha_x = \alpha_y$, it follows from figure 2.20 that for such a value of B the error in the determination of α is about 17%; hence $\alpha \simeq 19.5$ erg cm^{-2}.

As has been seen, the hillocks on the pyramid faces of these crystals have a triangular shape (figure 2.11). Kuznetsov et al (1987) measured $p(\sigma)$ of the steepest vicinal face of such a hillock (on an ADP crystal) produced by a single dislocation and obtained $\alpha \simeq 17$ erg cm^{-2}. This value is approximate because these authors used a simplified calculation scheme. They assumed that the time τ taken by a spiral to make one turn is

$$\tau = \sum_{i=1}^{3} l_{ci}/V_{\infty i}$$

where the index i refers to three vicinal faces. Then the slope steepness for $i = 1$ is

$$p_1 = h\sigma \left[\frac{gw}{k_B T} \left(\alpha_1 + \alpha_2 \frac{\beta_1}{\beta_2} + \alpha_3 \frac{\beta_1}{\beta_3} \right) \right]^{-1}.$$

Here β_i are the kinetic coefficients and g is the coefficient relating the dependence of the step segment velocity to the length. It was accepted that $g = 9.43/3\sqrt{3} = 1.82$, where $\lambda/l_c = 9.43$ was obtained by Budevski et al

(1975) for a triangular spiral under the assumption of the isotropy of α and β. For an ADP pyramid face $h = 5.33\,\text{Å}$. Kuznetsov *et al* (1987) found that $(p_1/\sigma)(k_B T/hgw) = 36.4\,\text{erg cm}^{-2}$, $\beta_i = 0.35,\ 0.5$ and $0.8\,\text{cm s}^{-1}$ for $i = 1,\ 2$ and 3, respectively. Hence,

$$\alpha_1 + \alpha_2(0.35/0.5) + \alpha_3(0.35/0.8) = 36.4\ \text{erg cm}^{-2}.$$

Assuming $\alpha \simeq \alpha_1 \simeq \alpha_2 \simeq \alpha_3$, these authors obtained the above mentioned value of α. A more exact value could be obtained using a technique for plotting the curves of figure 2.20 discussed in the previous section.

Let us now discuss the question as to what extent a local alteration of α at the point of the dislocation outcrop can affect the hillock periphery slope and the value of the surface free energy calculated by the dependence $p(\sigma)$. Such an alteration may occur as a result of the adsorption of an impurity, as well as due to strains in the crystalline lattice at a limited distance around the dislocation nucleus. The latter effect was studied by Cabrera and Levine (1956), Koziejowska and Sangwal (1985, 1988) and others.

At a distance ρ from the point of the dislocation outcrop, the allowance for the dislocation elastic energy enhances the crystal chemical potential by a value

$$\frac{Gb^2 w}{8\pi^2 \rho^2} = \frac{\rho_0 r_c}{\rho^2}\,\Delta\mu_\infty \tag{2.48}$$

where G is the Youngs modulus, b is the Burgers vector and ρ_0 is the Frank radius defined by the equation

$$\rho_0 = Gb^2/8\pi^2 \alpha. \tag{2.49}$$

ρ_0 has no physical meaning for small values of b and great α since it becomes smaller than atomic dimensions. However, with $G = 5 \times 10^{11}\,\text{erg cm}^{-3}$, $b = 10^{-7}\,\text{cm}$, $\alpha = 20\,\text{erg cm}^{-2}$ and $\rho_0 \simeq 3 \times 10^{-6}\,\text{cm}$ it was shown by Frank (1951) that a new hollow channel with radius ρ_0 should be created inside the dislocation. In this case in equation (2.49) α denotes the surface energy inside the channel. Allowing for equation (2.48) introduces a correction in the dependence of a step velocity on its curvature (see equation (2.7)), which now becomes

$$v = v_\infty(1 - r_c/r - \rho_0 r_c/\rho^2). \tag{2.50}$$

Cabrera and Levine (1956) showed that as a crystal grows, the dimensionless rate of rotation of an isotropic spiral ω_1 depends weakly on ρ_0. This rate decreases from 0.331 with $\rho_0 = 0$ to $\simeq 0.28$ with $\rho_0/r_c = 0.5-1$ and for greater values of ρ_0/r_c it remains constant. Far from the spiral centre the interstep distance remains $\lambda = 2\pi r_c/\omega_1$. Thus, allowing for equation (2.48) leads to an increase in the value of α calculated using $p(\sigma)$ by at most 15%. One can expect the error to be of the same order even if $\rho_0 = 0$ and a local alteration of α has occurred in the vicinity of the spiral centre with the adsorption of an impurity.

In dissolution the effects of dislocation elastic energy are much more significant. The solution saturated for a given crystal turns out to be supersaturated for a part of the surface in the spiral centre and the pit at the dislocation outcrop point begins to appear only with undersaturations greater than a certain critical value. According to the data obtained by Koziejowska and Sangwal (1985, 1988), the critical value of σ for a KDP prismatic face is -0.09, while for a pyramid face it is -0.064.

2.4.3   The kinetic coefficient

The kinetic coefficient β defined by equation (2.3) as the coefficient of proportionality between the step velocity and the supersaturation can be treated in different ways.

The conventional approach involves the assumption that the equilibrium concentration is always established on a step and its velocity is limited by diffusion processes. Burton *et al* (1951) showed that the most likely situation occurs when the number of growth points (kinks) is so great that it is comparable with the number of atoms on the step and incorporation into the step proceeds faster than material is supplied. The rounded shapes of dislocation hillocks and their constituent steps indicate the great number of kinks. Assuming that there is no interaction between kinks, in other words, that the attachment of a particle to a given kink is independent of the behaviour of particles at neighbouring kinks, we find that expression (2.3) reflects Fick's first law of diffusion as applied to the diffusion of substance towards a step.

Such a diffusion may occur directly from the bulk of the solution or it may be surface diffusion, as in the case of crystal growth from the gaseous phase where matter is supplied to a step by particles that have been previously adsorbed by the surface. When approaching the kink, the particles have to overcome an energy barrier, therefore

$$\beta = av \exp(-E/k_B T) \tag{2.51}$$

where a is the characteristic dimension of a building element and v is the frequency of oscillations ($v \simeq k_B T/h$, where h is Planck's constant). The activation energy E characterizes this barrier and refers to the desolvation processes of building elements and growth sites and to the corresponding steric and entropy factors.

The dependence of v on the amount of building matter per step must follow from the above considerations. In the case where the diffusion fields around neighbouring steps overlap each other, the step velocity must decrease. If the supply of matter owing to surface diffusion prevails, then when the interstep distance becomes less than $2\lambda_s$ (twice the free path of adsorbed particles across the surface) then v must become less than that corresponding to the linear law $v(\sigma)$. In other words, v must depend on p. A similar overlapping of the diffusion field must occur when matter is supplied onto a step directly from the bulk of the solution. In this case v will depend both on p and on the diffusion layer thickness, i.e. on the intensity of stirring of the solution. Hence, from the assumption that $c = c_0$ on the step, i.e. the supersaturation is always equal to zero, it follows that increasing the velocity of the solution flow (thus decreasing the diffusion layer width) must increase v however high the velocity of the flow.

It was mentioned at the beginning of this chapter that the growth kinetic regime always occurs at relatively low velocities of the solution flow and further increases in velocity do not affect R, p and v. Consequently, bulk diffusion does not limit v.

Experience shows that with actually achievable maximum values of p ($\simeq 10^{-2}$) the tangential growth rate is independent of hillock steepness. With such a hillock steepness as $\lambda = h/p \simeq 100h$ and if the free path $\lambda_s < \lambda/2 \simeq 50h$ ($\simeq 50$ parameters of an elementary cell), then surface diffusion does not limit the growth rate. In any case the fact that v is independent of p indicates unambiguously that under real conditions (when the solution has been sufficiently stirred) the growth rate is independent of diffusion processes, either bulk or surface ones.

Let us discuss in more detail the experimental evidence for this statement by considering prismatic faces of ADP, KDP and DKDP crystals.

In figures 2.21 and 2.23 one can see that $v(\sigma)$ remains linear over the entire interval σ, the values of v lying on the same straight line for different hillocks on the same face. During the experiment, p changed 2–3 times and reached a value of 5×10^{-3} (figure 2.23).

It should be noted that the measured dependences $v(\sigma)$ are seldom linear (for a prismatic face) as in the above figures. In the presence of impurities (including uncontrolled ones) the behaviour of this dependence is quite different: under low supersaturations the tangential velocity increases slowly, but beginning at a certain value of $\sigma = \sigma_*$ a steep rise is observed, while under still greater σ the dependence $v(\sigma)$ becomes a straight line passing through the origin. Such dependences are given in figures 2.27 (ADP), 2.28 (KDP) and 2.29 (DKDP). Without discussing at this stage the non-linear part of the curves $v(\sigma)$ which are to be treated in the next section, it should be noted that the straight line on which $v(\sigma)$ lies at sufficiently large values of σ is always the same regardless of the impurity content. It is this part of the curve that should be used to determine β, since in this region of supersaturations the effect of impurities turns out to be insignificant.

For the ADP prismatic face at $\simeq 33°C$ (figure 2.27) $\beta = 0.45\,\text{cm s}^{-1}$ ($w = 10.6 \times 10^{-23}\,\text{cm}^3$, $c_0^* = 2.049 \times 10^{21}\,\text{cm}^{-3}$, $v/\sigma = 9.7 \times 10^{-2}\,\text{cm s}^{-1}$).

For the KDP prismatic face at $\simeq 31°C$ (figure 2.28) $\beta = (7.8 \pm 1.7) \times 10^{-2}\,\text{cm s}^{-1}$ ($w = 9.68 \times 10^{-23}\,\text{cm}^3$, $c_0^* = 1.125 \times 10^{21}\,\text{cm}^{-3}$, $v/\sigma = (0.85 \pm 0.19) \times 10^{-2}\,\text{cm s}^{-1}$). At other temperatures (figures 2.30 and

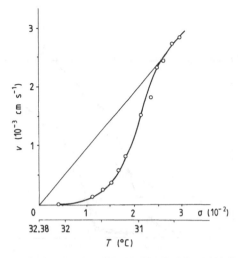

Figure 2.27 Supersaturation dependence of the tangential rate for the ADP prismatic face.

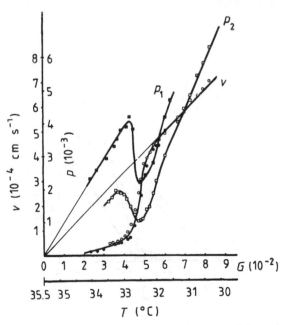

Figure 2.28 The supersaturation effect on the tangential rate and slope of two dislocation hillocks for the KDP prismatic face.

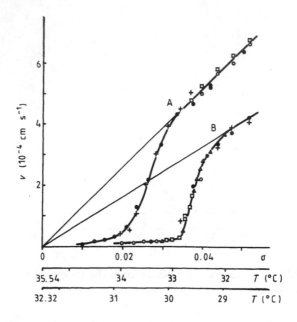

Figure 2.29   Dependence $v(\sigma)$ for the dislocation hillocks on the KDP prismatic face before (A) and after (B) addition of H_2O into the solution. The deuteration degree (%): A, 92; B, 85. Different dots on the curves refer to different hillocks. $T_{sat} = 35.54°C$ (curve A), $32.32°C$ (curve B).

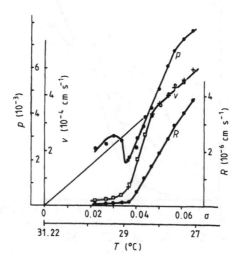

Figure 2.30   Growth characteristics of dislocation hillocks on the KDP prismatic face. Different dots on the curve of $v(\sigma)$ refer to different hillocks.

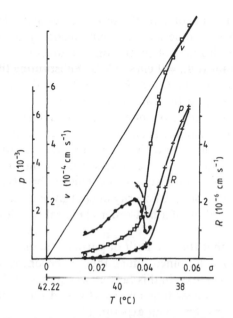

Figure 2.31 Growth characteristics of dislocation hillocks on the KDP prismatic face.

2.31) the kinetic coefficient of the steps turned out to equal: 28°C, $\beta = 7.2 \times 10^{-2}$ cm s^{-1} ($v/\sigma = 0.75 \times 10^{-2}$ cm s^{-1}); 38°C, $\beta = 12.2 \times 10^{-2}$ cm s^{-1} ($v/\sigma = 1.5 \times 10^{-2}$ cm s^{-1}). With the aid of equation (2.51) and making use of the latter two values for the activation energy we find $E = 9.8$ kcal mol^{-1}.

With this value of E at 33°C, β was found to equal 9.4×10^{-2} cm s^{-1} ($v/\sigma = 1.06 \times 10^{-2}$ cm s^{-1}), i.e. it is 5 times less than that for ADP at the same temperature regardless of the fact that the corresponding values of α are very close.

For DKDP (figure 2.29) with a deuteration degree $x = 0.92$ and at 33°C, β and v/σ are equal to 7.9×10^{-2} and 1.28×10^{-2} cm s^{-1}, respectively, while with $x = 0.86$, $\beta = 5.6 \times 10^{-2}$ cm s^{-1} and $v/\sigma = 0.81 \times 10^{-2}$ cm s^{-1}. Comparison with KDP shows that although the slope of the curves v/σ for DKDP is greater, their kinetic coefficient is less by about 20%.

All the above considerations refer to steps moving in the [001] direction.

For a hillock on the ADP pyramid face the kinetic coefficients for steps moving across the slopes of a pyramid hillock were measured by Kuznetsov *et al* (1987). On this face, v is usually linearly dependent on σ in the range of supersaturations studied. At $\simeq 35$°C in a solution of pH $= 4.4$ (the stoichiometric solution pH $\simeq 3.9$), $\beta_1 = 0.35$ cm s^{-1}, $\beta_2 = 0.5$ cm s^{-1}, and $\beta_3 = 0.8$ cm s^{-1}. The subscripts refer to the slopes labelled by numerals in figure 2.11. The values of β for the steps on the prismatic and pyramid faces are nearly equivalent.

It seems reasonable to compare the above values of β with the quite reliable and, apparently, still unique measurements made by Budevski *et al* (1980) for the electrocystallization of silver from an aqueous solution of $AgNO_3$. These authors obtained $\beta_{el} = 1 \text{ cm s}^{-1} \text{ V}^{-1}$. Substituting the supervoltage by supersaturation we obtain $\beta = \beta_{el} k_B T/e = 2.6 \times 10^{-2} \text{ cm s}^{-1}$ (*e* being the charge on an electron), i.e. this value is of the same order as that for KDP.

Returning to the discussion of the role of diffusion processes, we are inclined to conclude that the supersaturation near a linear step is close to that in the bulk solution and that it is not the introduction of building elements but their supply to the surface that is the slowest stage of the process. Defining β as a proportionality coefficient between v and the driving force ($\Delta\mu$) is still valid, but the linearity of $v(\sigma)$ given by equation (2.3) should be treated as a recording of a first-order chemical reaction proceeding on a step.

If it is assumed that the free path of particles across the surface $\lambda_s < 50h$ (an experimentally determined limit), then λ_s becomes comparable to the distance between the kinks. In this case, as was shown by Burton *et al* (1951), $v(\sigma)$ would not be a linear function, i.e. β would turn out to be dependent on σ, which is inconsistent with experiment.

2.5 The effect of impurities

It has been known for quite a long time that impurities, in particular ions of such metals as aluminium, iron and chromium, hinder the growth of the KDP-family crystals. Approximately 20–25 years ago, when the purification of starting materials was much less efficient, the prismatic faces did not grow at all; the lateral growth per year was 1–3 mm. Now the starting material contains less than 10^{-4} wt % of each impurity, and the normal growth rate in the [100] direction under sufficiently high supersaturation approaches that in the [001] direction. However, besides the above mentioned impurities there are other impurities whose nature is still unknown, which abruptly reduce the growth of a prism at low supersaturations, when at $\sigma < \sigma_*$ the face grows very slowly. The value of σ_* can reach 0.04 or more. Typical examples are given in figures 2.27–2.29. Similar curves are given also in figures 2.30 and 2.31 where one can see that with $\sigma \simeq 0.03$ the normal growth rate of a prism $R \lesssim 0.1$ mm per day, and at $\sigma \simeq 0.02 R < 0.01$ mm per day ($\simeq 1 \text{ Å s}^{-1}$). The appearance of σ_* may be connected with some organic impurities and a noticeable effect of the solution pH on the value of σ_* may be interpreted also as a manifestation of the effect of the dissolved salt hindering the growth. Unfortunately, these problems are still not understood.

Below we shall discuss first the experimental data on the chromium impurity and then some theoretical models

2.5.1 Some experimental results

The effect of the chromium impurity has been studied in many works: Torgesen and Jackson (1965), Byteva (1968), Davey and Mullin (1974a,b, 1976a,b), Dam *et al* (1984), Noor and Dam (1986), Bredikhin *et al* (1987) and many others. They established such facts as decreasing R increases the value of σ_*, the change in the surface morphology and habit of a crystal; however, the behaviour of $p(\sigma)$ and $v(\sigma)$ in the presence of chromium still remained unclear. The gap was bridged with the help of a technique described in section 2.1 (see figure 2.1, p 102).

Figure 2.32 represents the dependences $p(\sigma)$ and $v(\sigma)$ of the same dislocation hillock after addition of $CrCl_3 \cdot 10aq$ into the solution. One can see that the impurity Cr^{3+} causes p and v to decrease, the value of σ_* increasing from $\simeq 0.04$ to 0.05. In the presence of chromium a supersaturation dead zone appears: i.e. with $\sigma < \sigma_d < \sigma_*$ a crystal does not grow (Bredikhin *et al* 1987, Rashkovich and Shekunov 1990a,b). The value of σ_d increases with increasing concentration of chromium, and the difference $(\sigma_* - \sigma_d)$ decreases. With a concentration greater than about 6×10^{-6} mol Cr per mol KDP it is impossible to distinguish σ_* on the dependence $v(\sigma)$. Note also that upon addition of chromium, $\partial v / \partial \sigma$ increases along with the shift of the curves $v(\sigma)$ towards greater supersaturations near σ_*.

The minimum is observed on the curves $p(\sigma)$ with σ_*, here the hillocks change their shape as shown in figures 2.13 and 2.16. It is clear from the behaviour of the curves in figure 2.32 that if chromium is added to a solution of a given supersaturation when the hillocks have an elongated shape, then

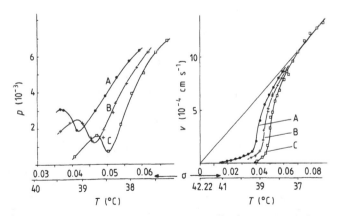

Figure 2.32 The effect of a chromium impurity on the slope p and tangential rate v of the dislocation hillock on the KDP prismatic face. The saturation temperature is 42.22°C. A, the initial solution (no added impurity); B, 3×10^{-6} mol Cr per mol KDP; C, 5×10^{-6} mol Cr per mol KDP.

first a decrease in p is observed and then the hillocks are transformed into elliptical ones. The latter process was also observed by Dam and van Enckevort (1984). One of the main peculiarities of the curves of $v(\sigma)$ is that they lie on one straight line passing through the origin with sufficiently high σ regardless of the concentration of chromium.

The concentrations of chromium employed (3×10^{-3} mol Cr per mol KDP corresponds to $\simeq 1.1 \times 10^{-4}$ wt % of KDP) are comparable to that in the salt which was used to prepare the solution (10^{-4}–10^{-5} wt %). If it is assumed that in a solution with no impurities the dependence $v(\sigma)$ is linear for all σ, then, since the shift of the curves $v(\sigma)$ is relatively small upon addition of chromium, one can conclude that in the initial solution there were also other impurities with an effect analogous to that of chromium.

Figure 2.33 shows a decrease in p, v and R of a hillock on the ADP prismatic face upon addition of chromium with $\sigma_d < \sigma < \sigma_*$. Figure 2.34 represents photographs illustrating the decreasing hillock steepness under the same conditions. Upon addition of the first portion of chromium a new steeper hillock appears on the face (frame A). Later on the steepness of both hillocks decreases, the new hillock completely devouring the old one.

In discussing the above facts one should first of all account for two phenomena: the decreasing tangential growth rate and the appearance of a minimum on the curves of $p(\sigma)$. Before taking up these questions, one more circumstance is worth noting. Decreasing v indicates that the impurity is quite firmly adsorbed by the surface or kinks. In this case the surface energy must decrease; however, with decreasing α the steepness of hillocks must increase not decrease as actually occurs (here we do not refer to the minimum

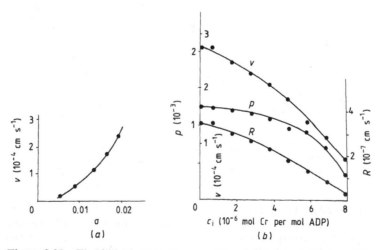

Figure 2.33 The hillock behaviour on the ADP prismatic face. The saturation temperature is 35.28°C. (*a*) Represents the dependence $v(\sigma)$ in the initial solution; (*b*) shows the effect of chromium with $\sigma = 0.021$.

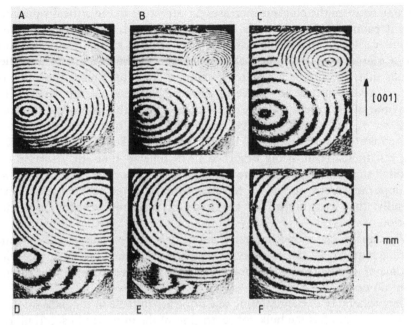

Figure 2.34 Decrease of the hillock slope with addition of chromium for the ADP prismatic face. The saturation temperature is 35.12°C and $\sigma = 0.033$. The experimental parameters are as follows (impurity concentration c_i in 10^{-5} mol Cr per mol ADP, time in minutes): A, (1.1, 0); B, (1.1, 4); C, (1.1, 10); D, (1.21; 20); E, (1.31, 25); F, (1.52, 75)

on the dependence $p(\sigma)$). This phenomenon is caused by the appearance of a dead zone: with $\sigma \leqslant \sigma_d$ there is no growth and $p = 0$. The dislocation generates a hillock only if $\sigma > \sigma_d$. Therefore in the interval $\sigma_d < \sigma < \sigma_*$ in the presence of chromium the curve $p(\sigma)$ lies lower than in a pure solution.

2.5.2 Tangential velocity

Usually two main mechanisms are considered that account for the hindering effect of impurities: the poisoning of the active sites of growth, i.e. kinks, by impurities (Bliznakov 1960); and the adsorption of impurity particles at a distance less than $2r_c$ hindering the advance of the steps (Cabrera and Vermilyea 1958).

If the lifetime of an impurity in the adsorption state is comparable with that of the crystal's own particles then, being adsorbed in the kink for a short time, the impurity reduces its active state and the velocity of the step without changing the shape of the latter. In this case the kinetic coefficient decreases but the linear dependence $v(\sigma)$ is preserved (Chernov 1961, 1984).

Blocking ions that persist in the adsorption state critically alter the behaviour of $v(\sigma)$. If the distance between them is less than $2r_c$, the step bending between the blockers increases its curvature to the critical value and then it cannot move any more. Such a situation is called the Cabrera–Vermilyea fence. With increasing σ the critical radius decreases and at $\sigma > \sigma_* = 2w\alpha\eta_s^{1/2}/k_BT$ the distance between the blockers $n_s^{-1/2}$ becomes greater that $2r_c$, and as a result the velocity $v(\sigma)$ approaches the value of that on a pure surface (n_s is the number of blockers per unit area). The behaviour of the experimental curves is consistent with this model; however, with $\sigma < \sigma_*$ the altered velocity, small as it is, is quite noticeable. It is reasonable to suppose that the lifetime of the blockers is not indefinitely long but finite. Then, being desorbed from time to time, they release the inhibited step which progresses to the next blocking ion. In the range $\sigma > \sigma_*$ the impurities seem, at first, to have no time to fasten to the surface and, secondly, they are trapped by the steps or repelled by them back into solution.

When analysing any mechanism of impurity inhibition one should compare the time required for the impurity to be adsorbed in the kink or surface with the time of 'exposure', i.e. the time taken by a new layer to cover the surface of an old one.

The probability of encountering one impurity particle in the bulk solution w is $c_i^* w$, where c_i^* is the bulk concentration of the impurity in solution. Let w be of the order of magnitude of a building element volume in the crystal, then $c_i^* w$ is the probability of encountering an impurity particle at the site of its adsorption. If adsorption requires the potential barrier E_i to be overcome, then the frequency of adsorptions occurring at random sites on the surface, i.e. the adsorption flow, will be

$$j = c_i^* w v \exp(-E_i/k_BT) \tag{2.52}$$

where $v \simeq 10^{12}\,\text{s}^{-1}$ is the effective frequency of attempts to overcome the adsorption barrier. During a time h/R ($= \lambda/v$ being the exposure of a terrace between the steps for adsorption) the fraction of sites occupied by the impurity from the total number of sites on the terrace will be $\vartheta = jh/R$. For the Cabrera–Vermilyea fence to be generated it is sufficient to cover the surface by blockers up to a value $\vartheta_* = h^2/\pi r_c^2$, where h^2 is the area of one adsorption site. Therefore, if the growth rates R are so high that $\vartheta < \vartheta_*$ or

$$R > R_* = j\pi r_c^2/h \tag{2.53}$$

then the impurity effect becomes insignificant since the impurity would have no time to become adsorbed. Calculating the value of E_i, at which this situation arises, as applied to the data given in figure 2.31 (the KDP prismatic face with $T \simeq 39°C$, $h \simeq 7 \times 10^{-8}$ cm, $R_* = 5 \times 10^{-7}$ cm s^{-1} at $\sigma = \sigma_* = 4.4 \times 10^{-2}$, $w \simeq 10^{-22}$ cm^3, $\alpha \simeq 20$ erg cm^{-2}). Substituting these values into equation (2.53) and taking (2.52) into account it is found that with $wc_i^* = 10^{-6}$, $R_* = 50 \exp(-E_i/k_BT)$ or $E_i \simeq 11.4$ kcal mol^{-1}. This value of the adsorption energy appears to be quite reasonable.

The model being discussed suggests that, v being independent of p, the exposure time for terraces on a steep hillock must be less, since the length of the terraces is less, and therefore the critical supersaturation σ_* for large p must be less than that for small p. Also, in the region of small $\sigma < \sigma_*$ the kinetic coefficient of steep hillocks must be greater since the impurity adsorption is less.

The experimental data reported above do not help to give a definite description of the behaviour of the curves $v(\sigma)$ with $\sigma \ll \sigma_*$ since for small R the measurements are lengthy and results are scarce. Nevertheless, the dependences $v(\sigma)$ in figures 2.21 and 2.23, where the lowest value of R measured was about 0.005 mm per day, can be interpreted as corresponding to the interval $0 < \sigma \ll \sigma_*$. In figure 2.21 the hillock steepness was in the range $\simeq (0.4-1.6) \times 10^{-3}$, $v/\sigma = (1.16 \pm 0.07) \times 10^{-3}\,\mathrm{cm\,s^{-1}}$, while for figure 2.23 the range was $2.5 \times 10^{-3} < p < 5 \times 10^{-3}$, $v/\sigma = (1.46 \pm 0.03) \times 10^{-3}\,\mathrm{cm\,s^{-1}}$. The corresponding values of β of steps progressing in the [001] direction were $(0.34 \pm 0.02) \times 10^{-2}\,\mathrm{cm\,s^{-1}}$ and $(0.43 \pm 0.01) \times 10^{-2}\,\mathrm{cm\,s^{-1}}$, which are about 100 times less than the value of β found for the second linear section of $v(\sigma)$ at the same temperature ($\beta = 0.47\,\mathrm{cm\,s^{-1}}$, see section 2.4.3, figure 2.27). The data given in figures 2.21 and 2.23 refer to a solution that had been used for a long time (nearly one year) and as a result had acquired impurities; no particular impurities had been added to it. These data seem to indicate an increase in β with increasing p at $\sigma \ll \sigma_*$. At the same time the experiments did not reveal any effect of the hillock steepness on the critical supersaturation σ_*, and consequently, the mechanism of the impurity effect is still not quite clear.

The Cabrera–Vermilyea fence model, however, makes it possible to explain some other facts which have not been mentioned (Rashkovich and Shekunov 1990a,b). Figure 2.35 represents the supersaturation dependences of the steepness and tangential velocity for two mutually perpendicular directions on one hillock marked by indices 2 and 1 (for both maximum and minimum v). Photographs of the hillock are shown in figure 2.36. In figure 2.35 one can see that the value of σ_* for $v_{\max}$ is less than that for $v_{\min}$. This fact gives rise to an abrupt change in the anisotropy of the hillock shape demonstrated by the curve $v_{\max}/v_{\min}$ constructed in the same figure. The greater the difference in σ_* for both directions, the higher the curve maximum must be.

Assuming that impurity blocking ions are evenly adsorbed on the whole surface and the surface energy of the step risers depends on their orientation and, accordingly, the lower the value of α, the smaller is the critical nucleus curvature radius, then it is reasonable to believe that the fence would be broken through first of all in the direction where r_c is lowest for a given value of σ. Taking into account that $r_c \sim \alpha/\sigma$, we obtain from equation (2.53)

$$(\alpha_1/\alpha_2)^2 = (R_{1*}/R_{2*})(\sigma_{1*}/\sigma_{2*})^2$$

where the subscripts refer to different directions and R_* is taken at $\sigma = \sigma_*$.

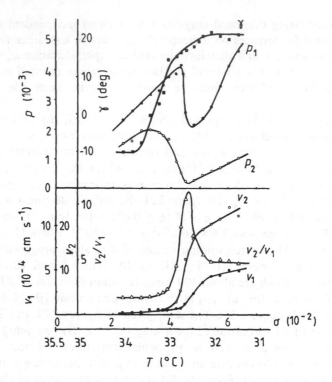

Figure 2.35 The effect of the supersaturation on the steepness and tangential rate for two directions on one hillock. γ is the angle between [100] and $v_{\max}$.

In view of these considerations the fact that σ_* is less for $v_{\max}$ indicates the inverse proportional relation between the kinetic coefficient and the surface energy of a step (if such a connection exists at all).

As applied to figure 2.35, α_1/α_2 took a value equal to 1.4–1.5 which practically coincides with the value v_2/v_1 at $\sigma \ll \sigma_*$ which represents the anisotropy of shape of the dislocation hillocks with small σ. Therefore one can expect that with $\sigma < \sigma_*$ the anisotropy of v is determined not only by the kinetic coefficient anisotropy but depends significantly on the anisotropy of α as well.

2.5.3   *Spirals with non-linear kinetics of steps*

To understand the reasons for the non-monotonous dependence $p(\sigma)$, Chernov and Rashkovich (1987a,b) considered the conditions of the progression of a spiral step with a non-linear dependence $v(\sigma)$ of the type observed in the experiments previously described (figure 2.35 and others). Since the spiral centre is fixed, the step curvature near it is equal to the critical value. The curvature diminishes with the distance from the centre, whereas the

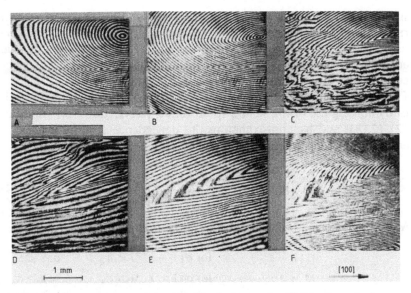

Figure 2.36 Change of the hillock shape on the KDP prismatic face. The growth characteristics of this hillock are shown in figure 2.35. $\sigma\,(10^{-2})$: A, 2.35; B, 3.9; C, 4.2; D, 4.5; E, 5.1; F, 6.35.

actual effective supersaturation (taking into account the Gibbs–Thomson effect) increases from zero in the spiral centre to σ on its periphery. Here the growth is supposed to proceed in the kinetic mode when the surface and bulk diffusion are not essential. If the step kinetics obey the linear law, i.e. $v \sim \sigma$, then the slower motion of the spiral step in the centre is due to the Gibbs–Thomson effect alone. In the case of non-linear $v(\sigma)$ the central parts of the spiral always progress much slower than the peripheral ones due to kinetic effects, in other words, no matter how strong the supersaturation (σ) is in the bulk of the solution, the central part of the spiral step always grows at supersaturations as low as $\sigma < \sigma_*$. At a certain distance from the centre the slow kinetics typical of $\sigma < \sigma_*$ is replaced by the fast kinetics associated with $\sigma > \sigma_*$. At the same time the period of the complete revolution of the spiral and hence the normal growth rate of the face is determined by the spiral centre. Under stationary conditions the normal growth rate $R = pv$ is the same both in the centre and on the periphery of the spiral. If we assume that in the centre $v = b_1\sigma$ and on the periphery $v = b_2\sigma$, where b_1 and b_2 are the kinetic coefficients for $\sigma \ll \sigma_*$ and $\sigma \gg \sigma_*$, respectively, then the slope at the periphery is $p = (b_1/b_2)p_0$, where p_0 is the hillock slope in the vicinity of the point of emergence of the dislocation. Thus, if $\sigma < \sigma_*$, the hillock slope is p_0. When the supersaturation in the solution bulk reaches σ_*, then the slope of the actually observed part of the hillock drops sharply (becoming b_2/b_1 times less steep). With further increases in the supersaturation the central part of the spiral (where the kinetic coefficient is small) becomes narrower (and drops to zero at $\sigma \to \infty$), and almost the whole spiral may be described by the linear dependence $v = b_2 \sigma$. The hillock slope is again equal to p_0. If the hillock is generated by a single dislocation, p_0 depends linearly on the supersaturation.

The above qualitative considerations can be transformed into quantitative ones by making use of the square spiral model (see section 2.3.1, equations (2.12)–(2.18)) and by approximating the dependence $v(\sigma)$ by a piecewise linear discontinuous function

$$v = \begin{cases} b_1 \sigma & \text{for } \sigma \leqslant \sigma_* \\ b_2 \sigma & \text{for } \sigma \geqslant \sigma_*. \end{cases} \tag{2.54}$$

Accordingly, instead of equation (2.12) the following dependence is used

$$u(s) \equiv v/b_2 \sigma = \begin{cases} r(1 - s^{-1}) & \text{for } \sigma(1 - s^{-1}) < \sigma_* & \text{or } s < s_* \\ 1 - s^{-1} & \text{for } \sigma(1 - s^{-1}) > \sigma_* & \text{or } s > s_* \end{cases} \tag{2.55}$$

where $r = b_1/b_2 \ll 1$ and $s_* = \sigma/(\sigma - \sigma_*) > 1$.

The kinetic coefficient of the step at one or several central segments is b_1 and at the other segments it is b_2. The transition occurs at a certain segment s_i which moves for some time $0 < t < t_*$ with kinetic coefficient b_1, and during a time $t_* < t < \tau$ its kinetic coefficient is b_2. Here the notation is the same as in section 2.3; the time t is measured in units of $l_c/b_2 \sigma$ from the moment at which the length of the zeroth segment (see figure 2.18) is $s_0 = 0$ and starts to extend. The instant t_* (at which the kinetic coefficient of the transition of the ith segment changes) and the number, i, of the segment itself are determined by the condition $s_i(t_*) = s_*$. If the supersaturation $\sigma > \sigma_*$ increases, then s_* and t_* decrease down to zero. For still higher supersaturations, the $(i - 1)$th segment becomes the transitional one and so on. Conversely, if the supersaturation decreases so that $\sigma \to \sigma_*$, then we have $s_* \to \infty$; in other words, the transitional segment occurs further removed from the centre. It is evident that each segment is transitional within a certain supersaturation range. The most extended interval including $\sigma \to \infty$ refers to the first progressing segment, $i = 1$. The transition from one branch of $v(\sigma)$ to another occurs in this segment until its maximum length $s_{1t} > s_*$. One can see from equation (2.58) that for $s_* \gg 1$ this condition is valid within the supersaturation range $\sigma_*(1 + r) < \sigma < \infty$, i.e. almost everywhere with $\sigma > \sigma_*$ and $r \ll 1$. Thus it is possible to restrict the calculations to this particular interval of supersaturations.

Thus, for $i = 3, 4, \ldots$, the solution of the form of equation (2.15) is valid. Equation (2.13) for ds_2/dt is integrated in the approximation that $s_1^{-1} + s_3^{-1} = 2s_2^{-1}$ and $1 - s_3^{-1} = 1 - s_2^{-1}$, which yields the following expression instead of equation (2.15)

$$2\tau = F(s_{2\tau}) + F(s_{2*}) - 2F(s_{1\tau}). \qquad (2.56)$$

The relation between s_2 and s_1 follows with the same accuracy from the equation

$$ds_2/ds_1 = \begin{cases} 1 & \text{for } 0 < t < t_* \\ 2 & \text{for } t_* < t < \tau \end{cases}$$

whence $s_{2\tau} = 3s_{1\tau} - s_* - 1$, $s_{2*} \equiv s_2(t_*) = s_{1\tau} + s_* - 1$ and

$$s_2 = \begin{cases} s_1 + s_{1\tau} - 1 & \text{for } 0 < t < t_* \\ 2s_1 + s_{1\tau} - s_* - 1 & \text{for } t_* < t < \tau. \end{cases}$$

Substituting $s_{2\tau}$ and s_2 into equation (2.56) gives

$$\tau = s_{1\tau} - 1 + 0.5 \ln[(3s_{1\tau} - s_* - 2)(s_{1\tau} + s_* - 2)/(s_{1\tau} - 1)^2]. \qquad (2.57)$$

The equation for s_1 follows from equation (2.17), if equation (2.55) and the above relation between s_2 and s_1 are taken into account:

$$s_{1\tau} - s_* - 1 - \frac{s_{1\tau} - s_* - 1}{s_{1\tau} - s_* - 2} \ln\left(\frac{s_{1\tau}}{s_*}\right) + \frac{s_{1\tau} - s_*}{2(s_{1\tau} - s_* - 2)} \ln\left(\frac{3s_{1\tau} - s_* - 2}{s_{1\tau} + s_* - 2}\right)$$

$$+ r\left[s_* - 1 + \frac{s_{1\tau} - 1}{s_{1\tau} - 2} \ln\left(\frac{s_{1\tau} + s_* - 2}{s_*(s_{1\tau} - 1)}\right)\right] = 0. \qquad (2.58)$$

Finding the root of equation (2.58) and substituting it into (2.57) the function $p(\sigma) = h/4\tau l_c$ can be determined. It may be conveniently represented in the form of the classical dependence $p_0(\sigma) = h/9.54l_c$, which is shown in figure 2.37 for $r \ll 1$.

Note that for $1 \ll s_* \ll r^{-1}$ we have

$$s_{1\tau} = s_* + 1 \qquad \tau = s_* + \ln 2. \qquad (2.59)$$

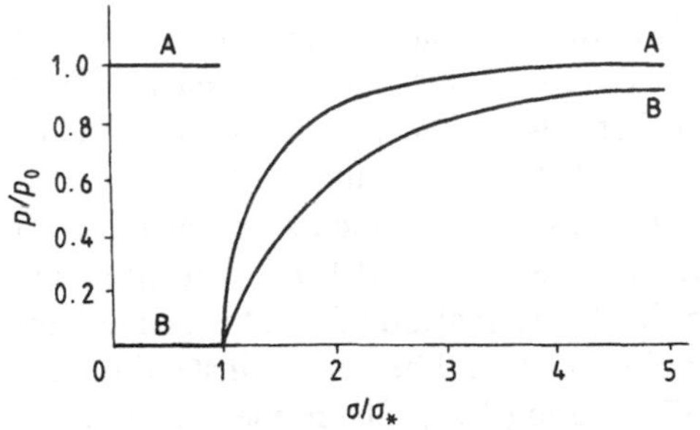

Figure 2.37 The supersaturation effect on the dislocation hillock steepness for the dependence $v(\sigma)$ of the type of equation (2.54) with $b_1/b_2 \to 0$ (curve A) and the same for $b_1 \equiv 0$ (curve B). $p_0(\sigma)$ corresponds to the linear $v(\sigma)$.

In the range $1 < \sigma/\sigma_* < 3$ the approximate relations (2.59) yield the slope p which is greater by at most 15% than the result of the exact calculations equations (2.58) and (2.57).

Mikhailov *et al* (1989) calculated $p(\sigma)$ for the case of non-linear dependence $v(\sigma)$ of the form of equation (2.54) for a circular spiral making use of equation (2.31). The result obtained differs only slightly from that given in figure 2.37.

Now let us consider qualitatively the anisotropic case. Let indices x and y denote the directions of maximum and minimum rates. Then under stationary conditions with $\sigma > \sigma_*$ the hillock slope in the centre p_0 and on the periphery would, as in the previous case, be connected by relations following from the constancy of the face's normal growth rate

$$p_x/p_{0x} = b_{1x}/b_{2x} = r_x \qquad p_y/p_{0y} = b_{1y}/b_{2y} = r_y.$$

Thus, the relative minimum depths on the dependence $p(\sigma)$ would be different for the two directions if the values of r_x and r_y differ. This is just the case shown in figures 2.35 and 2.36. In figure 2.35 it can be seen that $r_x < r_y$ and, as one might expect, the valley on the dependence $p(\sigma)$ is noticeably less than that for the minimum rate direction.

Experience shows that for the same dependence $v(\sigma)$ the values of the minimum of the curve $p(\sigma)$ differ greatly for various hillocks. The shape of hillocks also changes in a different manner, which is clearly seen by comparing the photographs in figures 2.36 and 2.38. To understand these phenomena one should consider the effect of the non-linear kinetics of steps on the shape of a spiral generated by a single dislocation.

The reader is reminded that in discussing the step source activity in section 2.4 it was shown that for large values of σ when $19r_c \ll 2L$ the hillock steepness no longer depends on the supersaturation. Therefore, qualitatively, it is clear that if σ_* happens to be within this supersaturation range, there would be no minimum on the dependence $p(\sigma)$ at all. The minimum depth for a multiple source, therefore, would be greater, the greater the source perimeter and the critical supersaturation value.

Quantitative calculations are easy to perform for the square spiral model where m dislocations are on the sides of a square of length L (in l_c units). Here the difference from the above calculations (formulae (2.12)–(2.18) and (2.55)–(2.58)) is that $s_0(\tau) = 1 + L$ and $s_0(0) = 1 = s_0(\tau)$ (see figure 2.39). For dimensionless lengths of all segments except the zero one the previous relation (2.14) still holds: $s_i(\tau) = s_{i+1}(0)$. This variation conserves the validity of the formula for the determination of τ but it generalizes the equations for the determination of s_{i_c}: L should be written instead of zero in the right-hand side of equations (2.18) and (2.58). The numerical solution to these modified equations with different values of s_* demonstrates that the dependences $\tau(L)$ obtained with $s_* = $ constant (i.e. with $\sigma/\sigma_* = $ constant) are nearly linear, at least to an accuracy of $\simeq 5\%$ they can be divided into 2–3 linear sections for $0 \leqslant L < \infty$. In other words, over sufficiently large intervals of L,

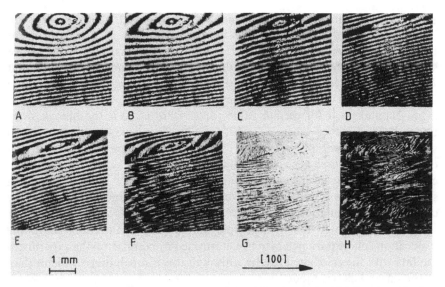

Figure 2.38 The change of the shape of the hillock on the KDP prismatic face. $\sigma(10^{-2})$; A, 3.65; B, 3.95; C, 4.26; D, 4.9; E, 5.2; F and G, 6.1; H, 9.3. Photographs G and F were taken at the same time, the latter with the reference mirror covered; one can see that the whole face is covered with macrosteps.

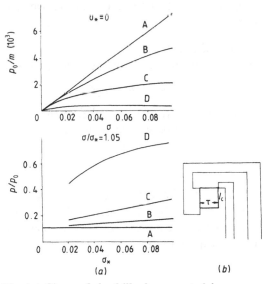

Figure 2.39 (*a*) Slope of the hillock generated by group dislocations with linear (p_0) and non-linear (p) kinetics of steps as $r \to 0$. $\hat{L}$ (μm): A, 0; B, 0.01; C, 0.05; D, 0.5. (*b*) Consecutive positions of the spiral segments, when the spiral goes round a square obstacle or a dislocation source.

$\tau = \tau_0 + kL$, where τ_0 and k depend on s_* only and $\tau_0 = \tau(0)$ over all intervals except the first one. This representation of $\tau(L)$ is convenient because in this case we have $\lambda = 4\tau l_c = 4(\tau_0 l_c + \hat{L})$, where $4\hat{L} = 4Ll_c$ is the source perimeter having the dimensions of length (it should be recalled that L is a dimensionless length). The results obtained are represented in figure 2.39 in the form of the dependences p/p_0 versus σ_* with $\sigma/\sigma_* = 1.05$ ($s_* = 21$) and $r \ll 1$ for certain values of $\hat{L}$. Here $p_0(\sigma)$ is the hillock slope under linear kinetics when $\sigma_* = 0$ ($s_* = 1$). The dependence $p_0(\sigma)/m$ is shown in the upper part of figure 2.39(a) for the same values of $\hat{L}$. To construct the curves it was assumed that $l_c = 2w\alpha/k_B T\sigma = 10^{-7}/\sigma$ cm, which corresponds to the surface energy value $\alpha \simeq 20\,\mathrm{erg\,cm}^{-2}$ at 35°C. One can see from the figure that the above qualitative considerations are fully confirmed by calculation.

It should be noted that in principle the above considerations allow the dislocation source parameters to be determined by comparing the experimental data with the calculated results, although this is much more difficult than in the linear kinetics mode.

Finally, let us consider one more question closely connected with the effect of impurities on the dependence $v(\sigma)$, i.e. how the dependence $p(\sigma)$ would behave if the dependences $v(\sigma)$ were different from those discussed in this chapter. Following Mickailov *et al* (1989) some types of $v(\sigma)$ are postulated which are to some extent similar to the experimental ones and we shall analyse the behaviour of a spiral taken in the natural form, $k(l)$ (see formulae (2.26)–(2.33), section 2.3.2).

2.5.3.1 The power dependence $v(\sigma)$

$v(\sigma)$ leads to a power relation between the step rate and its curvature:

$$v = b\sigma^n(1 - k/k_c)^n. \qquad (2.60)$$

First, let us discuss the qualitative differences from the classical case where $n = 1$. From equation (2.31) it follows that with $l \to 0$, $dv/dl = \omega$ is approximately valid. Whence $v = \omega l$ and consequently

$$k/k_c \simeq 1 - (\omega l/b\sigma^n)^{1/n}.$$

Within the dependence (2.60) the curvature diminishes with distance from the centre faster or slower depending on whether the value of n is greater or less than unity. Passing on to the dimensionless variables of equation (2.32) we obtain the following expression instead of equation (2.33) for the spiral equation:

$$y \int_0^x y(1 - y)^n \, dx - n(1 - y)^{n-1} \, dy/dx = \omega_1. \qquad (2.61)$$

The numerical calculation with $n = 1.5$ yields $\omega_1 = 0.252$ and with $n = 2$, $\omega_1 = 0.203$. Accordingly, at a great distance from the centre, $\lambda = 2\pi\omega_1/k_c =$

$24.93r_c$ and $30.95r_c$. It should be recalled that, with $n = 1$, $\omega_1 = 0.331$ and $\lambda = 18.98r_c$. The hillock slope is still proportional to the supersaturation and $R = pv \sim \sigma^{1+n}$.

2.5.3.2 Linear dependence with a dead zone

$$v = \begin{cases} 0 & \sigma < \sigma_* \\ b(\sigma - \sigma_*) & \sigma > \sigma_*. \end{cases} \quad (2.62)$$

Since, according to the Gibbs–Thomson rule, $\sigma = \sigma_0(1 - k/k_c)$, where σ_0 is the supersaturation at $k = 0$ and $k_*/k_c = (\sigma_0 - \sigma_*)/\sigma_0$, then $\sigma - \sigma_* = (\sigma_0 - \sigma_*)(1 - k/k_*)$. Omitting the subscript in σ_0 we obtain

$$v = \begin{cases} 0 & \sigma < \sigma_* \\ b(\sigma - \sigma_*)(1 - k/k_*) & \sigma > \sigma_*. \end{cases}$$

The value of k_* now plays the role of a critical curvature, therefore the lengths l and k^{-1} in equation (2.31) are normalized by multiplying by k_* instead of k_c: with $l = 0$, $k = k_*$; moreover, $\omega_1 = v/b(\sigma - \sigma_*)k_*$. As before, $\omega_1 = 0.331$ and $\lambda = 18.98k_*^{-1} = 18.98w\alpha/k_BT(\sigma - \sigma_*)$.

2.5.3.3 Linear discontinuous dependence with a dead zone

$$v = \begin{cases} 0 & \sigma < \sigma_* \\ b\sigma & \sigma > \sigma_*. \end{cases} \quad (2.63)$$

Contrary to the previous case, the curvature dependence of the velocity is still a classical one (see equation (2.7a)) for $\sigma > \sigma_*$ and the spiral shape satisfies equation (2.33). However, under the different boundary conditions at the spiral centre, for $l = 0$, $k = k_*$; i.e. $y(0) = k_*/k_c$. With $\sigma = \sigma_*$, $y(0) = 0$, and with $\sigma/\sigma_* \to \infty$, $y(0) \to 1$. This implies that a solution of equation (2.33) should be sought for each value of σ/σ_*. The corresponding numerical results are given in figure 2.37 (curve B).

The spiral equation in the form of equation (2.31) allows $p(\sigma)$ to be calculated not only for a discrete approximation as in equation (2.54) but also for the actual measured dependence $v(\sigma)$ represented in figure 2.35 and others. It is convenient to employ the following approximation

$$v(\sigma) = [b_1 + (b_2 - b_1)\{1 + \exp[d(1 - \sigma/\sigma_*)]\}^{-1}]\sigma. \quad (2.64)$$

As $\sigma \to \infty$, equation (2.64) yields $v = b_2\sigma$, while for $\sigma \to 0$, $v \to b_1\sigma$; the latter to an accuracy of up to $[(b_2 - b_1)/b_1]\exp(-d)$. Using an appropriately chosen constant, d, such an approximation describes the experimental data quite well.

In order to solve equation (2.31), (2.64) should be taken in the form

$$v = [r + (1 - r)\{1 + \exp[d(\sigma/\sigma_*)(1 - (\sigma_*/\sigma) - y)]\}^{-1}]b_2\sigma$$

where $y = k/k_c$ and $r = b_1/b_2$ as before, and $\omega_1 = \omega/b_2\sigma k_c$ should be introduced. As in the case of equation (2.63) the value of ω_1 must be found for each value of $0 \leqslant \sigma/\sigma_* < \infty$.

2.6   The stability of the growing surface

The equidistance of a train of steps generated by a dislocation source is often not conserved. Fluctuations in the velocity of separate steps result in the coalescence of elementary steps into larger ones. Under certain conditions this process may either be enhanced or diminished. The theory of the morphological stability of the growing surface has not yet been developed, primarily because of the lack of experimental data which makes the construction of an adequate model difficult. Therefore we shall confine ourselves to the consideration of individual factors responsible for the production of macrosteps.

The step section velocity v depends on three basic parameters: the supersaturation σ, the concentration of growth-retarding impurities c_i, and the orientation of a section under study φ. Local variation of each of these parameters results in a change in velocity; the greater the value of a corresponding partial derivative of v, the greater the change in the velocity. The greater the change in the velocity and hence the closer the steps, i.e. the greater the value of p, the more probable the coalescence of adjacent steps is.

2.6.1   Hydrodynamics effects

In the mixed kinetics regime the surface supersaturation is determined by the balance between the rate of transfer ot the crystallizing matter through the diffusion boundary layer due to molecular and convective diffusion and the rate of connection of the crystallizing matter into a crystal. At a low solution flow velocity, u, the diffusion layer is thinner at the front and at the lateral (with respect to the flow) edges of a face than above the middle, and, accordingly, the supersaturation is higher. Therefore the velocity of the steps adjacent to those edges is higher and the step bends. Ultimately, at the front edge the sides of the same step move toward each other and produce a pit. This process is illustrated in figure 2.40. The shape of the resulting pit depends on the number and arrangement of the dislocation hillocks on the face. In figure 2.41 a similar phenomenon is shown, but here at the front edge the steps generated by two dislocation hillocks coalesce.

In figures 2.40 and 2.41 one can see another phenomenon: the surface loses stability and macrosteps are produced at the back edge (with respect to the solution flow) on those hillock slopes where the steps progress in the direction of the solution flow.

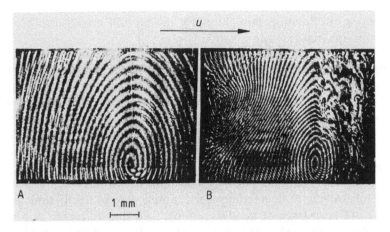

Figure 2.40 The formation of a pit and macrosteps on the ADP prismatic face. The solution flow velocity is $u = 15\,\mathrm{cm\,s^{-1}}$. $\sigma\,(10^{-2})$: A, 1.3; B, 2.1.

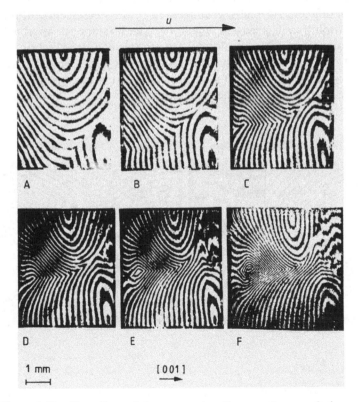

Figure 2.41 The effect of the supersaturation on the morphology of the ADP prismatic face in the mixed kinetics growth regime with $u = 15\,\mathrm{cm\,s^{-1}}$. $\sigma\,(10^{-2})$: A, 1.5; B, 1.8; C, 2.1; D, 2.3; E, 2.6; F, 2.7.

Reversing the solution flow direction, the pit and macrosteps exchange their positions (figure 2.42).

The production of macrosteps in a train progressing in the flow direction can be accounted for in the following way. When the solution moves along the face, the supersaturation gradient appears on it; the further from the front edge, the lower the value of σ is. For the steps progressing in the direction of decreasing values of σ, the probability of coalescence is greater since the front step moves more slowly than the back one. However, macrosteps of this type appear even at such flow velocities when the gradient of σ is small and does not affect the shape and steepness of the hillocks. Chernov *et al* (1986a) showed that if sections of various steepness appear on the hillock due to fluctuations (for example, due to fluctuations of σ at the point of emergence of the dislocation), then this difference in p would spontaneously decrease at the slope facing the flow and increase where the steps are progressing along the flow. Now let us consider the generating lines of such a hillock. Before perturbation they became piecewise linear. At the slope facing the flow, the solution approaches the convex joint point of sections of different steepness when passing by the steep section of the slope; when the solution has passed under the gentle slope, it approaches the concave joint point. Since the normal growth rate of a steep section is greater than that of a gentle one, then the solution at the convex point is relatively depleted, while at the concave point it is enriched with crystallizing matter. As a result, the roughness of the slope is smoothed out. At the opposite slope

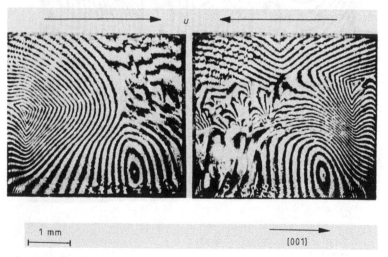

Figure 2.42 The effect of the solution flow direction on the morphology of the ADP prismatic face. The two photographs are separated by less than one minute after the change of the flow direction. $u = 15 \, \text{cm s}^{-1}$, $\sigma = 0.02$.

the situation is different: the solution approaches the convex point after passing under the gentle slope and it approaches the concave point after passing under the steep slope, therefore the disturbance of the relief increases with time.

As the dimensions of the face increase, the supersaturation at the emergence points of the growth sources changes and consequently the conditions of their competition are different. Changes in the activity of the sources often lead to transformations in the face relief: the bending of steps, the position and depth of a pit, and the intensity of macrosteps can be changed (Chernov *et al* 1986a, 1988, Rashkovich and Shekunov 1990a,b).

In figures 2.40 and 2.41 one can see that the development of the described phenomena depends strongly on the supersaturation in the solution volume. For low σ (and, accordingly, low R) the growth mode is close to the kinetic one even at low values of the velocity of the solution flow (see Chapter 3), therefore the supersaturation gradient at the surface is small; it increases with increasing σ. However, it is not only the supersaturation gradient that is essential, but the supersaturation region where the gradient takes place is also important. A pit and macrosteps on the prismatic face are most pronounced in the interval of σ corresponding to the steepest section of the dependence $v(\sigma)$. At the pyramid face $\partial v/\partial \sigma$ is constant under all values of σ, therefore the development of instability at this face is determined only by the gradient of σ at the surface.

Comparing the dependence $v(\sigma)$ for the prismatic faces of ADP and KDP indicates that the value of $\partial v/\partial \sigma$ for KDP is nearly 10 times less than that for ADP (see figures 2.27 and 2.28). Therefore at the KDP prismatic face the hydrodynamics effects are much less pronounced and, for example, a pit is not formed.

As the velocity of the solution flow is increased the supersaturation at the surface approaches the bulk supersaturation and the gradient of σ decreases. With increasing u, therefore, first a pit and then macrosteps disappear (the latter occurs at moderate values of σ). Figure 2.43 shows the disappearance of macrosteps when u is increased from 25 to 75 cm s^{-1}. The pit and the deformation of the steps disappear at still lower values of u. At small values of the flow velocity ($u \simeq 1$ cm s^{-1}) and at the horizontal position of the face a pit and macrosteps also do not appear. This is due to a small gradient of σ caused by a relative reduction of the supersaturation near the edges due to convective transfer of the solution depleted at the lateral faces.

Bending of the front of the steps and the formation of pits at the surfaces and macrosteps are quite common when a crystal is grown in solution; these effects are dangerous for the quality of a growing crystal. To avoid them one should choose an optimal interval of supersaturations (by relating it to the dependence $v(\sigma)$), and reverse stirring of the solution should be provided for. With the solution flow direction being changed periodically these defects have no time to form (Chernov *et al* 1986a).

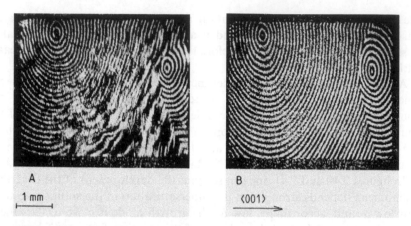

Figure 2.43 Disappearance of the macrosteps on the ADP prismatic face with increasing solution flow velocity for $\sigma = 0.021$. u (cm s^{-1}): A, 25; B, 75.

2.6.2 Macrosteps due to impurities

On increasing the supersaturation near σ_* the macrosteps (shown in figure 2.44) appear and then vanish. A distinctive feature of such macrosteps is that the velocity and direction of the solution flow do not affect the formation and intensity of these macrosteps (Rashkovich and Shekunov 1990a,b). These two facts make it possible to suppose that macrosteps appear when elementary steps pass through the fence of impurity blockers. Actually it turns out that the greater the amount of impurities in the solution and, accordingly, the greater the value of σ_*, the more intensive the macrosteps are. In solutions of KDP the value of σ_* is usually considerably greater than that in solutions of ADP. Therefore it is more convenient to study the

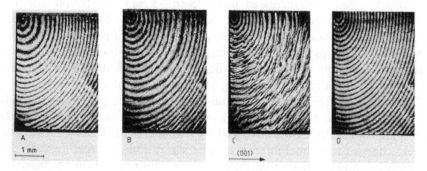

Figure 2.44 Formation of impurity macrosteps on the ADP prismatic face in the narrow range of supersaturations. $\sigma(10^{-2})$: A, 0.83; B, 1.24; C, 1.49; D, 1.83.

formation and evolution of macrosteps on the KDP prismatic faces. Besides, in ADP the change in the shape of a dislocation hillock takes place under supersaturations corresponding to the beginning of a linear section of the dependence $v(\sigma)$ after the disappearance of the macrosteps under study. In KDP both processes proceed at the same time, the change in the hillock shape being more important than in the case of ADP.

Figure 2.45 shows the formation of macrosteps and the change in the hillock shape on the KDP prismatic face. The macrosteps begin to develop in the direction of maximum velocity v. A region covered with macrosteps has a shape of sectors with a vertex at the hillock centre. When constant supersaturation has been established, the intensity of the macrosteps first increases and then diminishes (frames A and B). A further increase in σ leads to the broadening of the sectors covered with macrosteps (frame C). However, even now the intensity of macrosteps decreases. The emerging macrosteps are oriented mainly in the direction of maximum velocity (which is clearly seen in figure 2.45 in the lower row of photographs), which is normal to the movement of the elementary steps. Under still higher values of σ the surface becomes progressively and symmetrically covered with macrosteps of a different type (frame D, also see figure 2.14) which are enhanced with increasing supersaturation.

The following explanation of the process described above seems to be quite reasonable. At $\sigma \simeq \sigma_*$ an elementary step that had broken through impurity blockers elongates in the direction of maximum velocity (the minimum surface energy of a step riser) by forming individual protrusions (bulges) due to the statistically non-uniform distribution of the adsorbed impurities. Behind this step, the impurity concentration is lower, which initiates a filling of the interbulge space by the next step, and so on. The interstep distance in the direction of maximum velocity proves to be much larger than in the direction of minimum velocity, or, in other words, the slopes of the sides on the bulges are steeper and look like macrosteps. In the interference patterns they are revealed as spikes on the fringes (see figure 2.15 and frame C in figure 2.36).

With the change in σ the macrosteps start growing on the surface formed under the previous supersaturation. For the new value of σ, the orientation of elementary steps does not correspond to the previous anisotropy v (figure 2.35), thus provoking morphological instability. After a sufficiently long exposure the surface will be covered with steps generated by the source, in accordance with the new anisotropy of the tangential velocity, and the effect of this factor ceases. In this way it is possible to explain the maximum intensity of macrosteps in time under constant supersaturation.

The mechanism of the formation of macrosteps parallel to the elementary ones (as in the case of ADP, see figure 2.44) seems to be different. This can be clearly seen in the range of σ where the anisotropy in v is small. A sudden deceleration of the step that has broken through the step-blocking ions

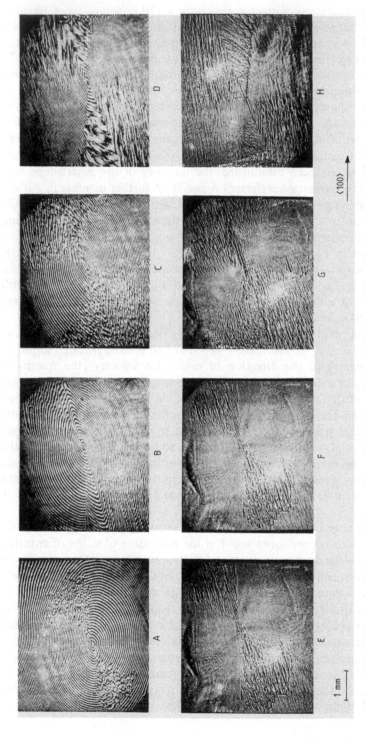

Figure 2.45 Change in the morphology of the KDP prismatic face with increasing supersaturation. The lower photographs (E–H) were taken with the interferometer reference mirror covered. σ (10^{-2}): A and B, 4.82; C, 5.13; D, 6.36. Photographs A and B were taken 1 min and 15 min after establishment of the constant value of σ. Photographs C and D were taken immediately after establishment of constant σ.

permits the next step to move along the impurity-free surface to catch the first step. The steps of different heights progress with different velocities; thin steps catch the thick ones, thus increasing their height and making the surface rough. This process should be most intensive in the region of σ where small fluctuations in the adsorbed impurity concentration give rise to the maximum changes in the tangential velocity. It is clear that the corresponding supersaturation range is in the vicinity of σ_*.

One should expect chromium to be adsorbed more readily at the surface than other uncontrolled impurities that are always present in a nominally pure solution; as a result a dead supersaturation zone is formed. It is found that the value of σ_d increases with the time spent by a crystal in the dead zone: it takes about 10 min for the surface chromium concentration to equal the volume concentration. As a result, with increasing supersaturation the cleansing of the surface from impurities has the important features demonstrated in figure 2.46.

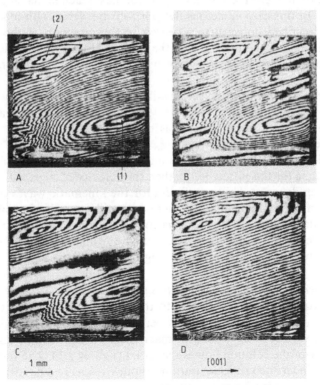

Figure 2.46 Cleansing of the chromium impurity-poisoned KDP prismatic face. $c_i = 9$ ppm. The crystal was kept in the dead zone for 12 hours with $\sigma = 0.078$. Time elapsed after establishment of constant τ (min): A, 1; B, 6; C, 20; D, 40.

A hillock (1) (shown in frame A of figure 2.46) was formed with $\sigma \gg \sigma_d$, and then the supersaturation was rapidly reduced to $\sigma < \sigma_d$ so that the hillock had no time to level out. On further increasing the supersaturation to $\sigma > \sigma_d$ the steps start moving near the edges. This is due to the fact that since the chromium distribution coefficient is less than unity, the impurity concentrates in the middle of the face; its concentration being less near the edges. A hillock (2) emerges on the surface region cleansed by the step motion. This hillock (2) in contrast to the poisoned one, (1), continues to grow. The lower steps of hillock (1) moving to the face periphery form a flat region, whereas the steps in the upper part of hillock (1) remain immobile, while the steps of hillock (2) continue propagating (frames B and C). Finally, hillock (2) absorbs the motionless hillock (1) (frame D). If hillock (2) were not formed or were less steep than hillock (1), then hillock (1) would have to start growing again when the flat portion reached its top.

Evidently, since hillock (1) remained motionless, σ_d was not reached. Attempts to determine σ_d by a very rapid increase in σ to make the hillock grow before the cleansed surface portion has reached its top were not successful. In this case numerous flat portions bordered with macrosteps of the type described above first appeared in the direction of the maximum tangential velocity and then all over the surface. However, the hillock started growing again only upon the cleansing of its summit by the nearest flat region.

The most general case for the possible formation of macrosteps or step clusters—kinematic waves (Chernov 1961, Cabrera and Coleman 1963)—under the influence of impurities has been studied in detail by van Erden and Müller-Krumbhaar (1986).

If the time of adsorption or the impurity lifetime is comparable with the exposure time for the surface (λ/v), then the velocity of steps will depend on the interstep distance. This is sufficient for the kinematic waves of the steps to be produced. Unfortunately, there are still no reliable data on the characteristic adsorption times for various impurities, and available experiments are few and inconclusive (Chernov and Malkin 1988). Therefore, it is important to perform the corresponding measurements.

2.6.3   Macrosteps forming under high supersaturation

At supersaturations corresponding to the linear portion of the $v(\sigma)$ curve, macrosteps are formed, the process being more intense for higher σ. The whole face surface is covered with macrosteps, which do not disappear as the velocity of the solution flow is increased (figures 2.14, 2.36 frame F, 2.45 frame D). Under such supersaturations impurities have no deceleration effect and, perhaps, a third factor is responsible for the production of macrosteps. This factor is connected with the polygonization of dislocation hillocks on the prismatic faces of KDP and DKDP crystals and always comes into play on the polygonized hillocks on the dipyramidal faces of ADP, KDP and

DKDP crystals. In these cases the hillock shape is determined by the segments of the steps moving in the directions of minimum tangential velocity. All other orientations of the step segments are not realized since for them v is greater and these segments are tapering. Since other orientations, however slightly they may differ from the above mentioned ones, do not appear, one can conclude that the function $v(\varphi)$ has a sufficiently deep minimum. Therefore, a fluctuational variation of the orientation of a step section should lead to a drastic increase in its velocity. If the steps are close enough, which is the case for the high supersaturations under study, this section may catch up with a neighbouring step thus increasing its thickness. There are no apparent reasons for such disturbances to relax. It should be pointed out that with $\sigma \geqslant 0.1$ the faces of crystals with polygonized hillocks are always covered with macrosteps which is a situation difficult to avoid.

3 Growing Single Crystals

Single crystals of the KDP family have been grown for more than 50 years (Loiacono 1987). In most cases, the linear dimensions of these crystals do not exceed 10 cm and the growth rate is not higher than 0.5–1 mm per day. It should be recalled that 40 years ago, when the quality of crystals was not as important as it is nowadays, the growth rate was quite high, for example Robinson (1949) succeeded in growing an ADP crystal 20 cm long at a rate of about 5 mm per day and Ansheles *et al* (1945) grew crystals of seignette salt at a very high rate.

Later the growth rate was steadily reduced because it was believed that this promotes a higher quality of crystal. In the last decade the demand for crystals has sharply increased, the requirements for their optical perfection have become stricter, and it has become necessary to achieve greater dimensions (Bordui 1987, Bespalov *et al* 1987a,b). Since a traditional technological growth cycle takes a long time, it is not feasible to grow crystals 20–30 cm long by increasing considerably the growth rate.

To achieve this aim two problems must be solved. The first one is connected with the fact that to increase the growth rate the saturation should also be increased, but this often leads to spontaneous crystallization which is difficult to avoid. The second problem is to maintain the high quality of the crystals at an increased growth rate. Considerable advances have been made in solving these problems and high-quality crystals with dimensions more than 20 cm have been grown at a rate of 20 mm per day (Bespalov and Katsman 1984, Rashkovich 1984). In the following, attention will be paid mainly to the growth of large optically perfect crystals at high rates.

3.1 Methods of creating supersaturation

The techniques used to grow crystals in solution differ mainly in the ways in which the supersaturation is created. Three such methods are known; the lowering of the temperature (for substances whose solubility increases with

166

temperature), the evaporation of solvent, and solution replenishment (Wilke 1963, Mullin 1972).

Technologically, the temperature reduction technique is the simplest and therefore the commonest method. Unfortunately, it is not suitable for growing large crystals since in this case a large volume of solution is required. For example, to grow a KDP crystal of dimensions $\simeq 20$ cm and weight $\simeq 20$ kg by the temperature reduction method more than 90 kg of solution is required. The rate of temperature reduction is usually chosen empirically. In this case care is taken to maintain a constant linear growth rate. At a high rate such a reduction of temperature is inadmissible since it results in considerable variations of the supersaturation. Since the temperature varies, in order to keep R constant, σ should be increased as the temperature is lowered. The corresponding calculations are not reliable because growth proceeds in the mixed kinetics regime where the role of the diffusion processes of the substance in the solution volume to the growing crystal surface is important and varies all the time.

For example, figure 3.1 shows the temperature dependences of supersaturations that provide constant growth rates (from 10 to 55 mm per day) of a

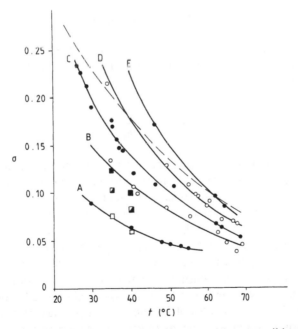

Figure 3.1 The experimental dependences $\sigma(t)$ responsible for the constant KDP growth rate in the [001] direction. R (mm per day): A, 10; B, 20; C, 30; D, 40; E, 50. The broken curve refers to the limiting supersaturation of the solution. □, ◪ and ■ represent the data of Mullin and Amatavivadhana (1967) at 10, 20 and 30 mm per day, respectively.

KDP crystal (Zaitseva 1989). The curves were obtained for the growth at a seed point in an extremely pure solution (less than 10^{-4} wt % metal ion impurities). The volume of the air-tight crystallizer was 5 l, the plate with the crystal was alternately rotated clockwise and anticlockwise at the rate of 60 rpm. To determine the supersaturation, the volume of a growing crystal was measured every hour; in these experiments the volume reached $\simeq 100$ cm^3. The curves were reproduced quite well since the seeds were regenerated under high supersaturations, and the activities of the dislocation sources responsible for the growth turned out to be practically the same. The reproducibility was worse if the seed regeneration was performed slowly under low supersaturations (<0.03). Mullin and Amatavivadhana (1967) and Mullin (1970) in experiments with the same supersaturations attained high growth rates, but the results agreed only for the rate of 10 mm per day (see figure 3.1). Those experiments, however, were made using crystals of $\simeq 2$ mm dimensions at a solution flow velocity of 15 cm s^{-1}, which is much greater than that corresponding to the data of figure 3.1. The behaviour of the curves in figure 3.1 can be understood from the following simple considerations. Writing a generally applied approximate formula for the dependence of the growth rate versus the supersaturation and temperature

$$R = b_1 \exp(-E/k_B T)\sigma^n \qquad (3.1)$$

where b_1 and n are constants, E is the activation energy for the kinetic coefficient, k_B is the Boltzmann constant and T is the absolute temperature. Then, with $R = \text{constant}$

$$d\sigma/dT = -\frac{E}{nk_B}\frac{\sigma}{T^2}.$$

Hence, the curves in figure 3.1 must have a negative slope which is enhanced with lowering temperature and increasing σ. Moreover, at a constant temperature the slope must increase with increasing σ; in other words, the greater the value of R, the steeper the curves of $\sigma(t)$ must be.

It should be noted that on the faces of crystals grown at high rates one can readily observe one or two dislocation hillocks surrounded by large macrosteps covering the whole surface. This fact indicates that even at $\sigma \simeq 0.3$ the dislocation growth mechanism dominates the two-dimensional nucleation. A similar situation was observed by Bespalov *et al* (1987a,b) for KDP and DKDP crystals grown at rates of 25 and 12 mm per day, respectively.

Evaporation of water as a means of creating supersaturation is used less often because this process is difficult to control at comparatively low temperatures (less than $\simeq 40°C$). Besides, to grow a large crystal, too much water has to be evaporated which results in a significant reduction of the solution volume. Thus, to grow a KDP crystal 20 kg in weight at 40°C,

$\simeq 60\,\text{kg}$ of water has to be evaporated; $\simeq 19\,\text{kg}$ are needed at $100\,°C$. Here we shall further consider the growth of KDP crystals from boiling solution (Rodriguez and Veintemillas 1984).

In these experiments a saturated solution ($c_0 \simeq 50\,\text{wt}\%$) was boiled at $\simeq 103\,°C$. A recycling refrigerator was used with the condensate being removed at a required rate. The growth rate was 4.3 mm per day with $\sigma \simeq 0.05$, i.e. it was less than the rate in non-boiling solutions (see figure 3.1). Mass crystallization began at $\sigma \simeq 0.09$. The comparatively poor stability of the supersaturated solution was apparently due to cavitation upon boiling. The x-ray topographic study showed that the crystals grew by a two-dimensional nucleation mechanism (Veintemillas *et al* 1987). The low growth rate can be supposed to be attributed to a high degree of polymerization of the building units (note that there are only four water molecules to each K^+ or $H_2PO_4^-$ ion). Thus, this method can hardly be used to grow large crystals.

Growing crystals by the replenishment method is considered to be the most promising technique. Its attractive features are that the temperature is kept constant during the growth process and the possibility of using a relatively small solution volume since crystallizing material is added as required. Two techniques of this kind aimed at growing large crystals deserve special attention.

Loiacono *et al* (1983) proceeded from the assumption that the deterioration of the properties of crystals grown at a rate higher than 1–2 mm per day is caused by fluctuations in the growth rate due to inadequate control over the supersaturation level. These authors devised a crystallizer running at a constant temperature and constant supersaturation. They showed that under these conditions the growth rate of crystals with dimensions of $\simeq 5\,\text{cm}$ could exceed 5 mm per day with no defects usually being observed. The set-up comprised three tanks—a crystallizer, a superheater and a dissolver—of a total volume of 144 l. The solution was in contact with Teflon and glass only. The temperature of each chamber was stabilized to an accuracy of $\pm 0.005\,°C$. The chambers were connected with heated tubes with valves. From the dissolver, the solution passed through a $100\,\mu m$ filter into the superheater and then it was pumped through a $0.6\,\mu m$ filter at a velocity of up to 7 l per hour into the crystallizer. From the crystallizer the solution passed into the dissolver. The supersaturation was determined from the temperature difference of the crystallizer and the dissolver which ranged from 0.5 to $5\,°C$. The temperature of the superheater was 10–$15\,°C$ higher than that of the crystallizer. The set-up turned out to be difficult in operation since the thermal and concentrational equilibrium took a long time to be established at the beginning of a technological cycle, and because of the hazards of spontaneous crystallization and the risk of the filter blocking up. Equilibrium was reached in 140–160 hours in the course of the experiment with varying temperature.

In experiments at $35\,°C$ and $\sigma \simeq 0.07$, the growth rate was $\simeq 5.5\,\text{mm}$ per day, i.e. slightly less than that implied by the data of figure 3.1. In addition, significant variations of the growth rate were observed which have been attributed by some authors to an inadequate flow of solution around the faces.

Yokotani *et al* (1984) suggested a simpler technique based on the replenishment method. In their experimental set-up (with the solution volume being about 2001) oppositely charged ions are transferred under the influence of an electric field through ion-selective membranes from a two-chamber dissolver into a crystallizer. KDP crystals with a cross section of 5 × 5 cm and up to 18 cm long were grown at a rate of up to 3–5 mm per day at a current of 1 A. The supersaturation was not measured. The system is comparatively simple in construction and operation, its disadvantages being the difficulty in stabilizing the supersaturation and the release of hydrogen and oxygen at the electrodes.

The advantages of isothermal growing are utilized when the solvent condensing on the crystallizer cover is used to dissolve new portions of the salt. One such system with a thermo-electrically cooled cover has recently been suggested by Vannay (1987).

The above methods of creating supersaturation have been implemented in various other crystallization set-ups. However, an optimal system for the rapid growth of large crystals does not seem to have been devised. An urgent task here is to create a reliable device to measure the supersaturation since its absence is one of the major obstacles to growing large crystals.

3.2   The stability of supersaturated solutions

In most cases when growing crystals, spontaneous nucleation occurs by the heterogeneous nucleation mechanism. A number of methods are known which are usually employed to reduce the number of points where nuclei may arise. These include the superheating of a solution in order to dissolve microcrystals (from solution and impurities), filtration to remove insoluble particles, the reduction of rough surfaces in the crystallizer, the prevention of cavitation by stirring the solution, and others. Degassing the solution is also a useful means to prevent nucleation (Owczarek and Wojchechowski 1987). It has also been established that the stability of solutions is reduced in the presence of ions (anions being tested mostly) of highly soluble impurities (Nagalingam *et al* 1980, 1981, Shanmungham *et al* 1984, 1985). Ions of heavy metals produce the same effect. It is well known that 15–20 years ago, when the raw materials for ADP and KDP crystals contained many more impurities than at the present time, it was far more difficult to preserve supersaturated solutions.

For many substances the maximum possible supersaturations, above which spontaneous crystallization begins, have been measured by experiment, and a number of empirical formulae have been derived which allow the value of the supersaturation to be calculated (Mullin 1972, Synowiec 1973, Mersmann and Förster 1984, etc). Those results, however, are not reproducible since they depend strongly on the purity of solution and experimental conditions.

The study of nucleation processes usually involves the determination of the time, τ, which is taken by the first nucleus to appear in solution at a given value of σ. The theory predicts that the dependence of $\ln \tau$ on σ^{-2} should be linear. By the slope of this curve one can calculate the free surface energy α which is the only fundamental parameter responsible for the rate of seed formation in a supersaturated solution at constant temperature. In many cases the above dependence becomes linear only under high super-saturations, for example, as seen in the experiments of Shanmungham *et al* (1984) with $\sigma > 1.5$. This fact can be interpreted as a result of the heterogeneous formation of seeds at low values of σ. Coincidentally, even with a linear dependence of $\ln \tau$ on σ^{-2} (Paul and Joshi 1976) one cannot be sure that a homogeneous process takes place since the values of α obtained are too low. Thus, for KDP crystals Joshi and Antoni (1979) obtained $\alpha = 2.4 \, \text{erg cm}^{-2}$, which is most likely to prove nucleation at the foreign surface. In similar experiments Zaitseva (1989) measured $\alpha \simeq 13 \, \text{erg cm}^{-2}$, which indicates, first of all, a thorough purification of the solution.

Figure 3.2 shows the data obtained in the latter study, which characterize the maximum attainable supersaturation in a well stirred solution of KDP with and without a growing crystal at various temperatures (the solution volume is 5 l). Zaitseva believes that the differences in the curves are caused by differences in the preparation of the solutions. The curve representing the data without a crystal was obtained under conditions where the filtered

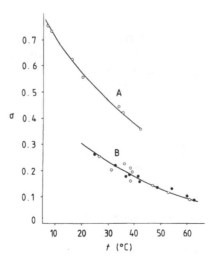

Figure 3.2 Temperature dependence of the highest supersaturation to be attained for the KDP solution. The curves correspond to $\tau = 1$ hour: A, without a crystal ($\bigcirc$); and B, with a growing crystal ($\bullet$).

solution was placed into the crystallizer and heated there to 80°C where it was kept for 3 days and was then cooled at the rate of 0.8°C per hour until the first nuclei appeared. In order to obtain the second curve the filtered solution was superheated in another tank, after which the solution was cooled to a temperature 2–3°C higher than the supersaturation temperature, and was then poured into the crystallizer containing a seed. The solution was further cooled at the same rate as in the first case. If an experiment without a crystal were to be conducted in this way, one would obtain the same curve as in the experiment with a crystal. These facts once again show how difficult it is to preserve the sterility of the solution. Curve B of figure 3.2 is plotted in figure 3.1 which shows that the growth rate of $\simeq 30$ mm per day may be realized without the hazard of spontaneous crystallization.

It has long been known that in the presence of growing crystals, nuclei appear usually at lower supersaturations than for the case where there is no crystal in the solution. The processes involved are known as secondary nucleation (Strickland-Constable 1968). Secondary nucleation is observed at quite low supersaturations when one or more small crystals appear in the crystallizer. The formation of secondary nuclei is caused by a number of factors and several models have been suggested that describe the process in detail (Wissing *et al* 1986, Blem and Ramanarayanan 1987, Kubota and Kubota 1986). The principal experimental evidence is the genetic connection between secondary nuclei and a growing crystal. This fact is manifested most distinctly when highly deuterated tetragonal DKDP crystals are grown. In this case small monoclinic crystals are formed by spontaneous crystallization whereas tetragonal ones are formed as a result of secondary crystallization (Zaitseva 1989).

In practice, to prevent secondary nucleation it is necessary to eliminate factors causing microparticles to break from a growing crystal. Special care should be taken to fasten a seed in such a way that when a crystal grows around glue, fasteners and plate roughnesses, no crystallizing pressure or inner stress caused by it should appear. Secondary nuclei often accompany the appearance of fractures and solution inclusions in a crystal. Enveloping the inclusions may be attended by fragments of growth layers covering the inclusion.

To conclude, curve B plotted in figure 3.2 characterizes the limiting supersaturation of a solution both with and without a growing crystal.

3.3 Achievement of the kinetic growth mode

Variation of the solution flow velocity with respect to the crystal faces may lead to changes in the normal growth rate of the faces (Ansheles *et al* 1945, Prieto and Amorós 1981, Takubo 1985). The connection between the step movement and the flow velocity was described by Robertson (1981). At the same time macrosteps may emerge on the surface (Rodriguez *et al* 1979, Chernov *et al* 1986b, 1988) and inclusions may be captured by the solution (Rosmalen and Bennema 1977, van Enckevort *et al* 1982). These effects are connected with changing the supersaturation at the crystallization surface σ_s, and its spatial and temporal inconstancy. With a well controlled bulk supersaturation, the choice of optimal hydrodynamic conditions in a crystallizer is a principal factor ensuring the high quality of crystals. The surface and bulk supersaturations should be made as close as possible in the kinetic growth regime so that the diffusion processes of supplying crystallizing matter do not limit the crystallization rate.

The crystallization regime criterion can be obtained by equating the material supply rate to the growth rate of a crystal.

The latter is expressed by equation (3.1) and so

$$b\sigma_s^n = k(\sigma - \sigma_s). \tag{3.2}$$

Here, the right-hand side of the equation represents the supply rate of the crystallizing material, and k is a function of both the solution flow velocity and certain material constants. We can rewrite equation (3.2) as

$$(\sigma_s/\sigma)^n = (k/b\sigma^{n-1})(1 - \sigma_s/\sigma) \tag{3.3}$$

and it becomes clear that the kinetic growth regime $(\sigma_s \to \sigma)$ occurs with

$$k/b \gg \sigma^{n-1}. \tag{3.4}$$

The inverse inequality leads to $\sigma_s \to 0$, i.e. to a purely diffusion regime. In the intermediate cases the mixed kinetic regime takes place. In particular, it follows from equation (3.4) that with $n > 1$ and with increasing σ, the crystallization is shifted into the diffusion region. Moreover, it may be seen from (3.4) that there are two ways to achieve the kinetic regime: k can be increased and b can be decreased. Let us consider the former case.

The right-hand side of equation (3.2) refers to Fick's first law of diffusion

$$k = Dc_0/\delta\rho$$

where D is the diffusion coefficient, ρ is the crystal density, and δ is the thickness of the diffusion layer, i.e. the solution layer where the main variation in the concentration occurs. The expression for δ can be obtained by the combined solution of the Navier–Stokes equation and that for convective diffusion (Levich 1962). In the particular case of a laminar flow around a semi-infinite thin plate

$$\delta = 3(D/v)^{1/3}(xv/u)^{1/2} \tag{3.5}$$

where v is the kinematic viscosity of the solution, u is the solution velocity far from the surface, and x is the distance from the plate edge. Formula (3.5) is considered to be valid for Reynolds numbers $Re < 10^3$. This expression is often used to estimate δ and σ_s when crystals are grown.

However, the real situation turns out to be different (Rashkovich and

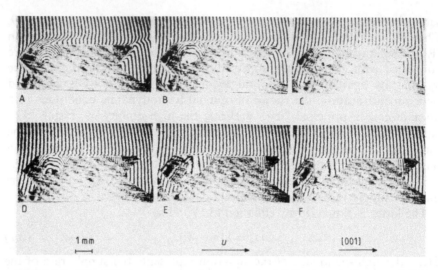

Figure 3.3   The diffusion boundary layer over the ADP prismatic face. $\sigma = 0.036$.
u (cm s^{-1}): A, 0.4; B, 1.5; C, 4; D, 6; E, 7; F, 15.

Shekunov 1990b, Rashkovich *et al* 1990). In figure 3.3 the interference patterns are given which characterize the change in the solution refractive index caused by changes in the solution concentration near the ADP prismatic face surface at different flow velocities. The shift of an interference line by one fringe corresponds to a decrease in σ by 0.003. The photographs help to determine the shape and thickness of the diffusion layer. At low values of u (frame A) the effect of natural convection is still strong, when the depleted solution moves upward. Frames B and C correspond to the laminar solution flow regime. With increasing velocity (frame D) the boundary layer breaks off. The upper bend of the interference fringes refers to the boundary layer break at the front edge, the lower bend reflects the concentration change within the reverse surface flow. It may be seen from frame D that in a flow circulating in this way (in a vortex hovering above the surface) there exists a region with a constant concentration that is nearly equal to the bulk concentration. Further increases in u lead to periodic separations of the boundary layer. One such case is shown in frame E. With increasing u the size of vortices diminishes and the frequency of their formation increases. With $u = 15$ cm s^{-1} (frame F), at a small distance from the front edge the flow is turbulent, the diffusion layer thickness becoming less than 10 μm and can only be discerned at the front edge. Note that under these conditions Re $\simeq 15$, and making use of equation (3.5) we obtain $\delta \simeq 20\,\mu$m.

In figure 3.4 the solution flow behaviour is visualized by the trajectories of bubbles produced in the electrolysis of the solution on two platinum wires which can be clearly seen in the figure. Frame A refers to the laminar flow

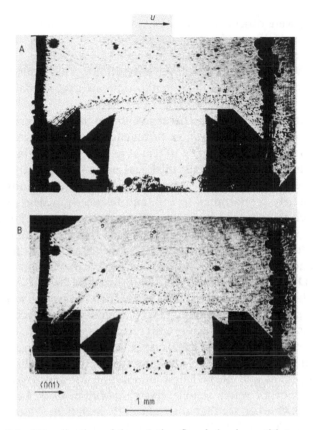

Figure 3.4 Visualization of the solution flow behaviour with tangential flow around the ADP prismatic face. u (cm s^{-1}): A, 3; B, 6.

regime, frame B displays a vortex hovering above the front half of the face. The trace lengths give information on the flow velocity. Almost stationary bubbles near the surface in frame B are in the reverse flow region.

A characteristic feature of figure 3.3 is that at the front edge the supersaturation is always less than that above the rest of the surface, although there the diffusion layer is thinner. This is associated with the fact that the pyramid face encountering the flow grows much faster (about 10 times faster as can be seen by comparing the dimensions of the crystal in frames A and F in figure 3.3) than the prismatic face. Therefore, the solution near the prismatic face is considerably depleted by the rapidly growing pyramid face.

Note that Rosmalen (1977) in experiments on the visualization of smoke flowing around a KDP crystal of the same geometry as shown in figures 3.3 and 3.4 observed vortex flows above the back (with respect to the flow) pyramid face with Re $\simeq 10^3$. At low Reynolds numbers no vortices above the pyramid were observed probably because it is difficult to realize a value of Re $\simeq$ 10 with the gas flow.

The above data indicate that at low $u \simeq 5\,\mathrm{cm\,s}^{-1}$ (Re $\simeq$ 10) and with a laminar flow regime, boundary layer separation occurs. This fact does not allow us to employ formula (3.5) even to estimate approximately the values of δ and σ_s.

If the solution flow is directed perpendicular to the face, then no separation of the boundary layer and no flow turbulence take place near that face, at least with u up to a value of $\simeq 100\,\mathrm{cm\,s}^{-1}$. The diffusion layer above the whole face has a nearly constant thickness being only slightly thinner near the edges (see figure 3.5). The layer thickness above the middle of the face is approximately proportional to u^{-1}, and for an equivalent value of u the layer is much thinner than in the case of tangential flow. For example, with $u > 5\,\mathrm{cm\,s}^{-1}$, $\delta < 10\,\mu\mathrm{m}$ and $\sigma_s \simeq \sigma$. Thus, these conditions are more favourable for the growth of crystals. This conclusion follows from the study of Prieto and Amorós (1981). As early as 1957 a patent was taken out for growing ADP crystals, in which it was suggested to direct the solution flow perpendicular to the growing pyramid faces. Then the growth rate was several centimetres per day (Wilke 1963, p. 124). This solution flow technique around a crystal has not been used again until recently.

The difference between the surface and bulk supersaturation can be assessed by the dependence $R(u)$. Figure 3.6 shows typical corresponding curves both for the tangential and normal flow of solution around a face, which are characteristic of all water soluble crystals (Mullin 1972, Lefaucheux and Robert 1979). One can see that at a sufficiently high flow velocity the growth rate becomes independent of the solution velocity and $\sigma_s = \sigma$. The function

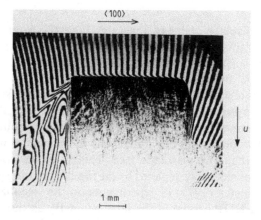

Figure 3.5 The diffusion boundary layer over the KDP prismatic face; the flow direction is normal to the face. $\sigma = 0.036$, $u = 3\,\mathrm{cm\,s}^{-1}$.

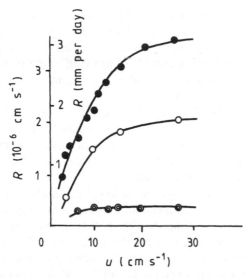

Figure 3.6 The growth rate dependence of the ADP prismatic face on the solution flow velocity. The curves were obtained for different but constant σ.

$R(u)$ reaches a plateau much faster than is expected from rough calculations (Chernov *et al* 1986b). This fact once again indicates the difficulty of determining σ_s by calculations for real conditions of crystal growth. The independence of R from U occurs at a certain value of U. This value of U is greater with increasing R. The corresponding function is nearly linear, i.e. to achieve the kinetic growth regime of an ADP prism with increasing R by 1 mm per day one should make a value of u 8 cm s^{-1} greater.

Thus, for crystallization to proceed in the kinetic regime at $R \simeq 20$ mm per day, quite high solution flow velocities are required. Such velocities are difficult to achieve when growing large crystals.

As far as the second possible way to satisfy condition (3.4), by means of reducing the kinetic coefficient b, one should take the following considerations into account. As mentioned above, even on faces with linear dimensions greater than 10 cm and $\sigma > 0.02$ 1–2 growth dislocation sources are found. The dependences $p(\sigma)$ indicate that growth sources contain far from one dislocation. Consequently, under the given conditions the source is not yet decomposed into single dislocations and with a linear dependence $v(\sigma)$ typical of pyramid faces the growth rate is defined by the expression

$$R = \frac{mh\beta\sigma^2}{19\alpha w/k_{\mathrm{B}}T + 2L\sigma}. \tag{3.6}$$

It follows from equation (3.6) that at a given supersaturation, R is minimum if m is small ($m > 1$) and L is large. There is a possibility of controlling the number of dislocations in the source when point seeds are used. If such a seed is embedded into a plate-crystal holder deeply enough, then most dislocations arising during regeneration would leave the crystal before it emerges from the recess and is capable of growing isometrically. In this way the value of m may be reduced considerably. No means of controlling the source perimeter is yet available.

For large σ, formula (3.6) is valid but is not applicable because of macrosteps covering the whole face. The tangential velocity of macrosteps depends on their height and is less than that of steps of elementary height. As yet no theory has been developed which would allow the normal growth rate of such a surface to be estimated.

Another possible way to increase the ratio k/b in equation (3.4) is associated with decreasing the growth temperature. Since the activation energy of the diffusion process is several times less than that of the surface processes, a decrease in temperature leads to a greater reduction of b as compared to k.

In order to establish the kinetic regime when growing crystals at high rates many workers attempt to create optimal conditions of solution stirring. Yokotani *et al* (1983) have grown KDP crystals at a rate of 3–5 mm per day by stirring the solution with a propeller stirrer rotating at a rate of 400 rpm. In their experiments a stirrer 7 cm in diameter rotated above a stationary crystal in a crystallizer 25 cm in diameter and 30 cm in height.

At a lower rate of rotation (300 rpm) solution inclusions appeared in the pyramid faces. In Zaitseva's experiments (see figure 3.1, Zaitseva 1989) growth rates were higher; however, no inclusions appeared. Unlike the experiments described above, here it was not a stirrer but a plate with a crystal that rotated (at 60 rpm with the direction of rotation being frequently changed).

The realization of monosectorial growth is a considerable technological achievement. In this case a crystal is placed in a mould so that only one face can grow. The geometrical simplicity of the growing surface makes it easy to control the solution movement (Bespalov and Katsman 1984). KDP and DKDP monocrystalline blocks of 21×21 cm in section were produced at a rate of about 20 mm per day (Bespalov *et al* 1987b). Cooper *et al* (1986) reported on the production of such blocks of 3–5 cm in section at a rate of 50 mm per day.

To conclude, even under the kinetic regime conditions the formation of macrosteps at high supersaturations cannot be ruled out completely. On the one hand, their emergence is perhaps not connected with the presence of local gradients of supersaturation at the surface, while on the other hand, when the diffusion layer is thin, fluctuations of temperature and supersaturation in the solution volume can be transferred to the surface much more easily since the time taken by the substance to diffuse (in other words, the heat transfer time) through the layer is proportional to δ^2/D and $\delta < 10\,\mu m$ and amounts to less than 1 s.

3.4 Problems of crystal quality

There is no unique correlation between increasing the growth rate and the crystal quality. With an increasing velocity of steps, the number of point defects apparently increases, since the number of attempts of each constituent element to find its place in a crystal is reduced. At the same time the crystal can absorb fewer impurities (and alien colloidal particles) at the faces due to the reduced exposure time for the interstep terraces. The appearance of solution inclusions is not directly connected with the growth rate, since it is defined by the surface morphology, the distorted shape of the steps, the formation of macrosteps, their mutual orientation, and so on. In the majority of cases dislocations emerge during the formation of solution inclusions, consequently their number is also not directly connected with the growth rate.

Typical defects in crystals and their classification are well known (Chernov 1984). Therefore we shall consider only such peculiarities of the defect structure that are typically produced during crystal growth at high rates.

3.4.1 Sectorial inhomogeneity

X-ray topographic examination of the ADP pyramid faces carried out by Smol'skii *et al* (1984, 1985) revealed differences in the lattice parameters for the crystal regions formed by different slopes of a polygonized dislocation hillock. This phenomenon is called vicinal sectoriality and is due to the differential capture of impurities by the segments of steps moving in different directions. These authors associate this with the effect of the step orientation on the number and structure of the kinks in the step. The steps of sector 1 of the dislocation hillock absorb the impurity in the most intense way (see figure 2.11), and the steps of sector 2 and 3 absorb it equally intensely, therefore the boundary between sector 2 and 3 is not distinct. As the velocity of the steps is increased the difference in the absorption of impurities by different sectors becomes more pronounced, therefore the trace of vicinal boundaries in the crystal bulk is revealed only at $\sigma > 0.02$. On the prismatic faces the orientation of the steps also affects the capture of impurities, but here when the shape of hillocks is elliptical, the orientation changes permanently. The lattice parameters are also permanently changed, therefore the stresses that arise in the crystal are not sufficient for the contrast to be shown in topographs. In the case of KDP and DKDP crystals, vicinal sectoriality on the prismatic faces is clearly revealed at sufficiently high values of σ after the polygonization of hillocks (Smol'skii and Zaitseva 1991). Vicinal sectoriality accounts, in particular, for the non-uniform distribution of an iron isotope in such crystals as observed by Belouet (1980).

At the join of steps generated by different dislocation hillocks the diffraction contrast is also observed; this is an intervicinal sectoriality. It is more pronounced the more distinct the difference in the orientation of the joining steps is.

Intervicinal sectoriality makes a certain contribution into the normal sectorial structure of crystals when in many cases the growth boundaries of separate faces are clearly seen. Internal stresses between the four faces of a pyramid are most pronounced. The intensity of such boundaries depends on the position of dislocation sources growing on adjacent faces. If steps of equal vicinal sectors join together at the edge, the internal stresses are minimum. Otherwise, they are larger and are enhanced with increases in the growth rate.

The intersectorial boundaries tracing the shift of edges between the faces of the pyramid and prism are less distinct. This is due to the fact that the steps of the vicinal sector 1 on the pyramid face are nearly parallel to the edge as well as to the step on the prism face moving in the [001] direction. The boundaries of growth sectors of prismatic faces are still less distinct. Even at high supersaturations they are less pronounced than the vicinal sectoriality on those faces.

3.4.2 Striation inhomogeneity

This type of defect is also connected with differences in the lattice parameters due to the varying content of impurities. The striation inhomogeneity arises when the supersaturation changes abruptly and is revealed in the form of layers with different refractive indices which are parallel to the faces. This kind of inhomogeneity is caused not so much by the growth rate but by the relative value of the supersaturation variation. From this point of view it appears that at a high rate of growth even greater absolute values of the solution temperature variations are admissible than those possible in the case of growth at tangential rates. It is clear that for a striation inhomogeneity to appear it is the temperature variations in the region where $\partial v / \partial \sigma$ is maximum that are most harmful.

3.4.3 Anomalous biaxiality

Tetragonal ADP, KDP and DKDP crystals are uniaxial. Due to the structure defects and related internal stresses, however, they reveal an anomalous biaxiality (Shamburov and Kucherova 1965). This value is characterized by the angle 2v between the optical axes and the orientation of the plane of the optical axes with respect to the x and y axes of a tetragonal crystal. The relation between the anomalous biaxiality and anomalous birefringence in the xy plane is given by the expression

$$\Delta n = n_x - n_y = (n_0 - n_e)v^2.$$

It is accepted that crystals of high quality have a value of $2v \simeq 5'$, those of good quality $2v \simeq 15'$, while for those of average quality $2v \simeq 30'$. For an ADP crystal the relative values are $\Delta n = 2 \times 10^{-8}$, 2×10^{-7} and 10^{-6}.

Usually, a growth sector of each face has a particular orientation in the plane of the optical axes. For sectors of a prismatic face it is perpendicular to the face. For sectors of pyramid faces three orientations are typical: (i) parallel to the edge between a prism and a pyramid; (ii) parallel to the intersectorial boundaries of a pyramid; and (iii) perpendicular to them (Belyustin *et al* 1972). A combination of these types of orientation is often observed.

Annealing of crystals may significantly reduce the value of the anomalous biaxiality. It is reduced most effectively in the growth sectors of a prism and, in the case of the first type of orientation, in the optical axes' plane of pyramidical growth sectors. For crystals of the third type of orientation, annealing is not effective. Along with reducing 2v in the pyramid growth sectors, annealing changes the orientation of the optical axes' plane from type (i) to type (ii). The average value of 2v at the sample cross section upon annealing for two days at $\simeq 150°C$ becomes twice as small. However, at the sample point where the conoscopic pattern, by which the anomalous biaxiality is determined, is strongly distorted (in such cases the biaxiality is considered to be due to dislocations) annealing does not change the biaxiality (Azarov *et al* 1983). But there is in addition a variation in the axiality: a disorientation of an acute bisection of the optical axes' angle. The variation in the axiality is revealed in the broadening of isogyres of a conoscopic pattern with the broadening of the light beam passing through the crystal. Sometimes the position of the optical axes' plane is difficult to fix, which indicates a variation in its orientation along the crystal length.

For crystals grown at high rates biaxiality of the order of 2v = 20–30' is typical, i.e. slightly greater than for crystals growing slowly. For electro-optical applications where a complete screening of the light using crossed polarizers is required, minimum biaxiality is important since the resulting birefringence causes chaotic field brightening. In non-linear optics during the transformation of laser light frequencies, the anomalous biaxiality leads to variations in the phase synchronism directions in various parts of a crystal. Because of the small angular width of the synchronism it may reduce the frequency transformation efficiency. Calculation and experiment show that an anomalous biaxiality of 2v $\simeq$ 20' affects insignificantly the generation efficiency of the second harmonic of neodymium lasers, and crystals can be used to create optical frequency doublers with an efficiency of up to 90%, and also for other applications (Amandosov *et al* 1983, Akhmanov *et al* 1984, Bespalov *et al* 1987a,b).

3.4.4 Optical transparency in the ultraviolet region of the spectrum

Transmission spectra for ADP and KDP crystals with no prismatic face growth sectors are well known. The transparency region (defined as a wavelength region in which the transmittance is more than 10% with a sample thickness of 1 mm) for ADP is 0.12–1.7 µm, while for KDP it is 0.25–1.7 µm.

A prismatic growth sector, in contrast to a pyramid, strongly absorbs light in the wavelength region of 0.2–0.3 μm (Bespalov *et al* 1982, Amandosov *et al* 1983). This is especially important for growing crystals at high rates when the supersaturation is sufficiently high since prismatic faces grow at a rate comparable to that of pyramid faces. It has been known for a long time that prismatic faces capture more iron, chromium and aluminium than pyramid faces do. For example, with a content of 0.84% Fe^{3+} in a KDP solution, 1250 ppm of iron was found in a prismatic sector, while 6.3 ppm was found in a pyramid sector. Accordingly, with $\lambda = 0.25$ μm the absorption index was $\simeq 2\,cm^{-1}$ and $\simeq 100\,cm^{-1}$ (Dieguez *et al* 1985). Although there are quite diverse literature data concerning the distribution coefficients for metal ions in a solution and crystal, which is apparently due to a difference in the structural perfection of crystals (Belouet 1980), nevertheless it is only Fe, Cr and Al that are characterized by a definite difference in the distribution coefficients for prismatic and pyramid sectors. However, it is rather difficult to explain the absorption of light by a heavy metal impurity, since the solutions used for growing crystals contain only a small quantity of such impurities ($\simeq 1$ ppm). Besides, typical absorption curves are quite smooth and no peaks referring to these ions (for iron: 0.215–0.270 μm) can be observed.

Another possible explanation is connected with the presence of usually uncontrolled organic impurities in the solution (Amandosov *et al* 1983). These may be microbes reproducing in solution (Yokotani *et al* 1986), and lubricant particles from the rotating parts of the crystallizer or products dissolved from the parts made-up of organic glass (e.g. the cover, stirrer and seed plate).

There is a distinct correlation between the solution absorption spectrum in the wavelength region under consideration and the absorption spectrum of a prismatic growth sector. This provides hope that the problem of purifying solutions will be solved. For example, to remove microbes the following methods are found to be effective: ultraviolet irradiation of the solution, bubbling oxygen through the solution, or adding 10–100 ppm of hydrogen peroxide (Nishida *et al* 1988).

Unfortunately, numerous investigations have not yet elucidated the reasons for the differential introduction of heavy metals into the faces of the prisms and pyramids. This also applies to organic impurities. It is not impossible that it is the intrinsic structural defects and not the impurities that are responsible for the absorption of light by a prismatic sector.

3.4.5 Laser strength of crystals

This qualitative characteristic of crystals is essential as applied to powerful non-linear optics. The effect of growth conditions and the structural perfection of KDP crystals on their strength under powerful laser radiation has been studied in detail (Cooper and Singleton 1986). It has been established in this and other works (Amandosov *et al* 1983, Akhmanov *et al* 1984, Bespalov *et al* 1987b) that an increase in the growth rate up to 30 mm per day does not reduce the crystal strength.

The study of the relationship between the density of dislocations in KDP crystals and their laser strength revealed the absence of any correlation between these parameters (Newkirk *et al* 1983). In one of the early works (Endert and Melle 1981), however, the strength has been shown to reduce at a density of dislocations at the surface greater than 200 per cm^{-2}. Most crystals, however, grown by both conventional and high-rate techniques have a density of dislocations much less than this value.

Strong internal stresses with an anomalous biaxiality $2v > 30'$, which cannot be removed by annealing, result in a reduction of the strength. Decreasing the anomalous biaxiality by annealing a crystal (for 1–4 days at $\simeq 140°C$) enhanced the laser strength (Swain *et al* 1982). However, with $2v \lesssim 20'$, no correlation has been observed between the optical strength and the anomalous biaxiality.

Heavy metal impurities reduce the laser strength. For example, an increase of the chromium concentration in solution from 5×10^{-5} to 6×10^{-3} wt % lowered the damage threshold by a factor of 10 (Azarov *et al* 1985).

Typical values of the optical strength of KDP crystals for $\lambda = 1.053 \mu m$ are 3–8 J cm^{-2} (3–5 GW cm^{-2}) for an impulse duration of 1 ns (Yokotani *et al* 1986); Nishida *et al* (1988) succeeded in raising this value to 20–22 J cm^{-2} by removing the microorganism colonies from the solution (the growth rate of the crystal was 3–5 mm per day). These authors believe that the colonies can grow into a crystal forming microscopic hillocks on the prismatic faces (Yokotani *et al* 1987). Singleton *et al* (1988) determined the laser strength and organic impurities in KDP crystals received from different suppliers and found a definite correlation between these parameters. They believe that organic molecules coalesce into mycelium-like structures captured by a crystal. Absorption of light by these structures results in the local cracking of crystals. Increasing the laser strength up to 20 J cm^{-2} is a great achievement which makes KDP crystals the most resistant of all known optically non-linear single crystals (Eimerl 1987b).

References

Adhav R S 1969 Elastic, piezoelectric and dielectric properties of rubidium dihydrogen phosphate *J. Appl. Phys.* **40** 2725–7

Agbalyan Yu G, Tatevosyan L S and Sharkhatunyan R O 1975 The possibility of determining the supersaturation of the solution in the growth of lithium iodate crystals *Sov. Phys.–Crystallogr.* **20** 549–50

Aguilo M and Woensdregt C F 1987 Theoretical growth on equilibrium forms of ADP ($NH_4H_2PO_4$) *J. Cryst. Growth* **83** 549–59

Akhmanov S A, Begishev I A, Gulamov A A, Erofeev E A, Zhdanov B V, Kuznetsov V I, Rashkovich L N and Usmanov T B 1984 High efficient parametric frequency conversion of light in large-aperture crystals grown by a fast method *Sov. J. Quantum Electron.* **14** 1145–6

Alexandrovskii A L, Golovanov A I and Rashkovich L N 1978 Optical and nonlinear optical properties of monoclinic DKDP crystal *Abstract, Proc. IX Conf. USSR on Coherent and Nonlinear Optics (Leningrad, June 1978)* vol I (Moscow: Nauka) p 182

Amandosov A T, Pashina Z S and Rashkovich L N 1983 Quality of ADP crystals formed by rapid growth from point-like seed *Sov. J. Quantum Electron.* **13** 271–2

Amandosov A T, Velichko I A and Rashkovich L N 1981 A high-temperature phase transition in $K(D_xH_{1-x})_2PO_4$ crystals *Sov. Phys.–Crystallogr.* **26** 231–3

Anosov V Ya and Pogodin S A 1947 *Principles of Physicochemical Analysis* (Moscow: Academy of Science USSR)

Ansheles O M, Tatarskii V B and Shternberg A A 1945 *Fast Growth of Crystals from Solution* (Leningrad)

Avanesyan G T 1988 On the growth kinetics of an elliptical vicinal hillock *Sov. Phys.–Crystallogr.* **33** 1492–4

Averbuch-Pouchot M T and Durif A 1985 Structure of a new form of rubidium dihydrogenphosphate, RbH_2PO_4 *Acta Crystallogr.* C **41** 665–7

Azarov V V, Atroshchenko L V, Danileiko Yu K, Zakharkin B I, Kolybaeva M I, Minaev Yu P, Nikolaev V N and Sidorin A V 1985 Influence of structure defects on the internal optical strength of KDP single crystals *Sov. J. Quantum Electron.* **15** 89–90

Azarov V V, Atroshchenko L V, Kolybaeva M I, Leonova E N and Selin S M 1983 The effect of annealing on optical homogeneity of potassium dihydrogenphosphate single crystals *Bull. Acad. Sci. USSR, Inorg. Mater.* **19** 164–5

Bacon G E and Pease R S 1953 A neutron diffraction study of potassium dihydrogen phosphate by Fourier synthesis *Proc. R. Soc.* A **220** 397–421

Baikov Yu M 1986 Nonequilibrium thermodynamics of hydrogen isotopes and the isotope effect in crystallization of alkaline earth dihydrogen phosphates *J. Phys. Chem. (USSR)* **60** 758–60

Balascio J F, Loiacono G M, Hayden S C and Idelson A L 1975 Growth and characterization of $Cs(D_{1-x}H_x)_2AsO_4$ from aqueous solutions *Mater. Res. Bull.* **10** 193–200

Baranov A I, Fedosyuk R M, Ivanov N R, Sandler V A, Shuvalov L A, Grigas J and Mizeris R 1987 Phase transitions in monoclinic RbD_2PO_4 *Ferroelectrics* **72** 59–66

Baranov A I, Shuvalov L A, Ryabkin V S and Rashkovich L N 1979 Critical behaviour of the dielectric constant in the uniaxial ferroelectric CsH_2PO_4 *Sov. Phys. Crystallogr.* **24** 300–2

Baranowski B, Friesel M and Lunden A 1986 New phase transitions in $CsHScO_4$, CsH_2PO_4, $RbHSO_4$, $RbHSeO_4$ and RbH_2PO_4 *Z. Naturf.* a **41** 981–2

—— 1987 Adsorption of water as a means to remove bulk metastability of $CsHSO_4$ and RbH_2PO_4 *Z. Naturf.* a **42** 565–71

Barkova L V and Lepeshkov I N 1966 Properties of aqueous solutions of KD_2PO_4 *Russ. J. Inorg. Chem.* **11** 1470–2

—— 1968 Potassium oxide–phosphorus pentoxide–deuterium oxide system at 25° *Russ. J. Inorg. Chem.* **13** 1432–6

Bärtschi P, Mattias B, Merz W and Scherrer P 1945 Eine neue, seignette-elektrische modifikation von rubidiumphosphat *Helv. Phys. Acta* **18** 240–2

Bates R G 1951 First dissociation constant of phosphoric acid from 0 °C to 60 °C; limitation of the electromotive force method for moderately strong acids *J. Res. NBS* **47** 127–34

Batyreva L A, Bespalov V I, Bredikhin V I, Galushkina G L, Ershov V P, Katsman V I, Kuznetsov S P, Lavrov L A, Novikov M A and Shvetsova N R 1981 Growth and investigation of optical single crystals for high-power laser systems *J. Cryst. Growth* **52** 832–6

Belouet C 1980 Growth and characterization of single crystals of KDP family *Prog. Cryst. Growth Charact.* **3** 121–56

Belouet C, Monnier M and Crouzier R 1975 Strong isotopic effects on the lattice parameters and stability of highly deuterated DKDP single crystals and related growth problems *J. Cryst. Growth* **30** 151–7

Belyustin A V, Stepanova N S and Fridman S S 1972 Effect of annealing on optical anomalies of potassium dihydrogen phosphate crystals *Bull. Acad. Sci. USSR, Inorg. Mater.* **8** 1624–6

Bennema P, Boon J, van Leuwen C and Gilmer G H 1973 Confrontation of the BCF theory and computer simulation experiments with measured (R, σ) curves *Krist. und Techn.* **8** 659–73

Berg E 1901 Über phosphate des rubidiums und cäsiums *Ber. Dtsch. Chem. Ges.* **34** 4181–5

Berg L G 1938a The solubility isotherm of the ternary system $K_2O–P_2O_5–H_2O$ at 25 °C *Bull. Acad. Sci. USSR, Ser. Chim.* N1 147–60

—— 1938b The solubility isotherm of the ternary system $K_2O–P_2O_5–H_2O$ at 50 °C *Bull. Acad. Sci. USSR, Ser. Chim.* N1 161–6

Bergman A G and Bochkarev P F 1938 Physiocochemical study of the equilibrium of the aqueous reciprocal system of potassium and ammonium nitrates, monophosphates and chlorides *Bull. Acad. Sci. USSR, Ser. Chim.* N1 237–66

Bespalov V I, Batyreva I A, Dmitrenko L A, Korolikhin V V, Kuznetsov S D and Novikov M A 1977 Investigation of the absorption of near infrared radiation in partly deuterated KDP and α-HIO$_3$ crystals *Sov. J. Quantum Electron.* 7 885–7

Bespalov V I, Bredikhin V I, Ershov V P, Katsman V I, Kiseleva N V and Kuznetsov S P 1982 Optical properties of KDP and DKDP crystals grown at high rate *Sov. J. Quantum Electron.* 12 1527–8

Bespalov V I, Bredikhin V I, Ershov V P, Katsman V I and Lavrov L A 1987a Fast growth of water-soluble crystals and fabrication of large-aperture light frequency converters *Bull. Acad. Sci. USSR, Ser. Phys.* 51 1354–60

—— 1987b KDP and DKDP crystals for nonlinear optics grown at high rate *J. Cryst. Growth* 82 776–8

Bespalov V I and Katsman V I 1984 Growth of large water-soluble crystals for laser optics *Bull. Acad. Sci. USSR* Institute N9 11–14

Blem K E and Ramanarayanan K A 1987 Generation and growth of secondary ammonium dihydrogen phosphate nuclei *Am. Inst. Chem. Eng. J.* 33 677–80

Blinc R, Burgar M, Čižikov S, Levstik A, Kadaba P and Shuvalov L A 1975 Dielectric properties of monoclinic KD$_2$PO$_4$ *Phys. Status Solidi* b 67 689–94

Bliznakov G 1960 Adsorption of impurities and the mechanism of crystal growth *Sov. Phys.–Crystallogr.* 4 134–40

Bobylev A V and Rashkovich L N 1980 An interference method of measuring the rates of crystal growth in solution, and its application to KH$_2$PO$_4$ *Sov. Phys.–Crystallogr.* 25 255–7

Bordui P 1987 Growth of large single crystals from aqueous solution: a review *J. Cryst. Growth* 85 199–205

Bordui P F, Herko S P and Kostecky G 1985 Computer-controlled crystal growth of KDP from aqueous solution *J. Cryst. Growth* 72 756–8

Bordui P F and Loiacono G M 1984 In-line bulk supersaturation measurement by electrical conductometry in KDP crystal growth from aqueous solution *J. Cryst. Growth* 67 168–72

Bozin S E and Smecerovic M 1983 Growth of ammonium dihydrogen phosphate microcrystals under nonstationary conditions *J. Cryst. Growth* 65 487–93

Brazhnik P K, Davydov V A and Mikhailov A S 1988 Kinematical approach to description of autowave processes in active media *Theor. Math. Phys.* 74 300–6

Bredikhin V I, Ershov V P, Korolikhin V V and Lizyakina V N 1987 Influence of impurities on the growth kinetics of KDP crystal *Sov. Phys.–Crystallogr.* 32 122–5

Brodskii A I 1952 *Chemistry of Isotopes* (Moscow: Academy of Science USSR)

Brosheer J C and Anderson J F 1946 The system ammonia–phosphoric acid–water at 75 °C *J. Am. Chem. Soc.* 68 902–4

Broul M, Nyvlt J and Söhnel O 1979 *Tabulky Rozpustnosti Anorganickych Latek ve Vode* (Prague: Academia)

Buchanan G H and Winner G B 1920 The solubility of mono and diammonium phosphate *Ind. Eng. Chem.* 12 448–51

Budevski E, Bostanov V and Staikov G 1980 Electrocrystallization *Ann. Rev. Mater. Sci.* 10 85–112

Budevski E, Staikov G and Bostanov V 1975 Form and step distance of polygonized growth spirals *J. Cryst. Growth* 29 316–20

Burton W K, Cabrera N and Frank F C 1951 The growth of crystals and the equilibrium structure of their surfaces *Phil. Trans. R. Soc.* A **243** 299–358

Busch G and Scherrer P 1935 Eine neue seignette-elektrische substanz *Naturwissenschaften* **53** 737

Bykova I N, Kuznetsova G P, Kolotilova V Yu and Stepin B D 1968 RbH_2PO_4–$RbCl$–H_2O and CsH_2PO_4–$CsCl$–H_2O systems *Russ. J. Inorg. Chem.* **13** 540–4

Byteva I M 1962 Effects of pH and crystal-holder speed on the growth of crystals of ammonium dihydrogen phosphate *Growth of Crystals* vol 3 (New York: Plenum) pp 213–16

—— 1965 Effect of pH of medium on the growth of ammonium dihydrogen phosphate crystals *Sov. Phys.–Crystallogr.* **10** 105–6

—— 1966 Effects of the pH on the shape of ammonium dihydrogen phosphate crystals *Growth of Crystals* vol 4 (New York: Plenum) pp. 16–19

—— 1968 Effects of pH on the growth of ADP crystals in the presence of Fe^{3+} and Cr^{3+} *Growth of Crystals* vol 5B (New York: Plenum) pp 26–32

Cabrera N and Coleman R V 1963 Theory of crystal growth from the vapor *The Art and Science of Growing Crystals* ed J J Gilman (New York: Wiley)

Cabrera N and Levine M M 1956 On the dislocation theory of evaporation of crystals *Phil. Mag.* Ser. 8 **1** 450–8

Cabrera N and Vermilyea D A 1958 The growth of crystals from solution *Growth and Perfection of Crystals* (New York: Wiley) pp 393–410

Cerreta M K and Berglund K A 1987 The structure of aqueous solutions of some dihydrogen orthophosphates by laser Raman spectroscopy *J. Cryst. Growth* **84** 577–88

Chernov A A 1961 Laminar-spiral growth of crystals *Sov. Phys. Usp.* **78** 277–331

—— 1971 Theory of the stability of face forms of crystals *Sov. Phys.–Crystallogr.* **16** 734–58

—— 1984 Crystal growth *Modern Crystallography vol 3 (Springer Series in Solid State Sciences 36)* (Berlin: Springer) pp 1–278

Chernov A A, Kuznetsov Yu G, Smol'skii I L and Rozhanskii V N 1986a Hydrodynamic effects in the process of crystallization of ADP crystals from aqueous solutions in the kinetic mode *Sov. Phys.–Crystallogr.* **31** 1193–2000

Chernov A A and Malkin A I 1988 Kinetics of impurity stoppers formation preventing the growth of (101) ADP faces in a liquid solution *Sov. Phys.–Crystallogr.* **33** 1487–91

Chernov A A and Rashkovich L N 1987a Spiral growth of crystals with nonlinear dependence of step rate on supersaturation; (100) faces of KH_2PO_4 group crystals in solution *Dokl. Akad. Nauk USSR* **295** 106–9

—— 1987b Spiral crystal growth with nonlinear dependence of step growth rate on supersaturation; the (100) faces of KH_2PO_4 crystals in aqueous solution *J. Cryst. Growth* **84** 389–93

Chernov A A, Rashkovich L N and Mkrtchan A A 1986b Solution growth kinetics and mechanism: prismatic face of ADP *J. Cryst. Growth* **74** 101–12

—— 1987 Optical interference investigation of surface processes of growth of crystals of KDP, DKDP and ADP *Sov. Phys.–Crystallogr.* **32** 432–42

Chernov A A, Rashkovich L N, Smol'skii I L, Kuznetsov Yu G, Mkrtchan A A and Malkin A I 1988 Growth of KDP-group crystals from solution *Growth of Crystals* vol 15 (New York: Plenum) pp 43–91

Chernov A A, Smol'skii I L, Parvov V F, Kuznetsov Yu G and Rozhanskii V N 1979 Study of the growth kinetics of ADP crystals from a solution by the in-situ method of X-ray topography *Dokl. Akad. Nauk USSR* **248** 356–8

Chomjakow K, Jaworowskaja S and Schirokich P 1933 Die lösungs und verdünnungswärmen von KDP und ADP *Z. Phys. Chem.* **167** 35–41

Cooper J P and Singleton M F 1986 Damage resistance and crystal growth of KDP *Proc. SPIE* **681** 6–9

Cooper J F, Singleton M F and Zundelevich J 1986 Rapid growth of potassium dihydrogen phosphate crystals in the form of rectangular sector blocks *Abstract VIII, Int. Congr. on Crystal Growth (York, England, July 1986)* POA1/147

Courtens E 1987 Mixed crystals of the KH_2PO_4 family *Ferroelectrics* **72** 229–44

Dam B, Bennema P and van Enckevort W J P 1986 The mechanism of tapering of KDP-type crystals *J. Cryst. Growth* **74** 118–28

Dam B, Polman E and van Enckevort W J P 1984 In situ observation of surface phenomena on {100} KDP related to growth kinetics and impurity action *Industrial Crystallization 84* ed S J Jancic and E J de Jong (Amsterdam: Elsevier) pp 97–102

Dam B and van Enckevort W J P 1984 In situ observation of surface phenomena on {100} and {101} potassium dihydrogen phosphate crystals *J. Cryst. Growth* **69** 306–16

Dauncey L A and Still J E 1952 An apparatus for the direct measurement of the saturation temperatures of solutions *J. Appl. Chem.* **2** 399–404

Davey R J and Mullin J W 1974a Growth of the {101} faces of ammonium dihydrogen phosphate crystals in the presence of ionic species *J. Cryst. Growth* **23** 89–94

—— 1974b Growth of the {100} faces of ammonium dihydrogen phosphate crystals in the presence of ionic species *J. Cryst. Growth* **26** 45–51

—— 1976a A mechanism for the habit modification of ammonium dihydrogen phosphate crystals in the presence of ionic species in aqueous solution *Krist. und Techn.* **11** 229–33

—— 1976b The effect of pH on the growth of the {100} faces of ADP crystals *Krist. und Techn.* **11** 625–8

Dieguez E, Cintas A, Hernandez P and Cabrera J M 1985 Ultraviolet absorption and growth bands in KDP *J. Cryst. Growth* **73** 193–5

Dmitrenko L A and Korolikhin V V 1975 Spectrophotometric determination of deuterium content in crystals of deuterated potassium dihydrogen phosphate and solutions used for growing them *Sov. Phys.–Crystallogr.* **20** 282–3

D'yakov V A, Koptsik V A, Lebedeva I V, Mishchenko A V and Rashkovich L N 1973 High-temperature phase transition in crystals of $Rb(D_xH_{1-x})_2PO_4$ *Sov. Phys.–Crystallogr.* **18** 769–72

Eimerl D 1987a Electro-optic, linear, and nonlinear optical properties of KDP and its isomorphs *Ferroelectrics* **72** 95–130

—— 1987b High average power harmonic generation *IEEE J. Quantum Electron.* **QE-23** 575–92

Endert H and Melle W 1981 Influence of dislocations in KDP crystals on the laser damage threshold *Cryst. Res. Technol.* **16** 815–9

Eysseltova J and Dirkse T P 1988 Alkali metal orthophosphates solubility *Data Ser.* **31** 168–277, 307–33

Fedotova O K and Tsekhanskaja Yu V 1984 Concentration dependence of partial molar volumes of components of binary aqueous solutions of potassium dihydrogen phosphate, ammonium dihydrogen phosphate, ammonium nitrate and ammonium sulfate at different temperatures *J. Phys. Chem. (USSR)* **58** 1530–2

Fellner-Feldegg H 1952 Strukturbestimmung von CsH_2PO_4 *Tscherm. Min. Petr.* **3** 37–44

Flatt R, Brunisholz G and Chapius-Gottreux S 1951 Contribution à l'etude du systéme quinaire $Ca^{++}-NH_4^+-H^+-NO_3^--PO_4^{---}-H_2O$. IV. Les systéms ternaires limites $Ca^{++}-H^+-PO_4^{---}-H_2O$ et $NH_4^+-H^+-PO_4^{---}-H_2O$ à 25 °C *Helv. Chim. Acta* **34** 683–91

Frank F C 1951 Capillary equilibria of dislocated crystals *Acta Crystallogr.* **4** 497–501

Franke V D, Punin Yu O, Mirenkova T F, Vorob'ev A S and Ivanova T Ya 1975 Effect of solution pH on the kinetics of potassium dihydrogen phosphate crystal growth *Vestn. Leningr. Univ. Geol. Geogr.* **N2** 146–9

Fukami T, Akahoshi Sh and Hukuda K 1986 Deuterium concentration of $H_2PO_4^-$ and NH_4^+ radicals in partially deuterated $NH_4H_2PO_4$ *J. Phys. Soc. Japan* **55** 3633–5

Gallagher P K 1976 A study of the thermal decomposition of some alkali metal dihydrogen phosphates and arsenates *Thermochim. Acta* **14** 131–9

Gary R, Bates R G and Robinson R A 1964 Second dissociation constant of deuteriophosphoric acid in deuterium oxide from 5 to 50° Standardization of a pD scale *J. Phys. Chem.* **68** 3806–9

Gavrilova I V and Kuznetsova L I 1966 Aspects of the growth of mono-crystals of potassium dihydrogen phosphate *Growth of Crystals* vol 4 (New York: Plenum) pp 69–72

Giordmaine J A 1962 Mixing of light beams in crystals *Phys. Rev. Lett.* **8** 19–20

Gupta L C, Rao U R K, Venkateswarlu K S and Wani B R 1980 Thermal stability of CsH_2PO_4 *Thermochim. Acta* **42** 85–90

Havránková H and Březina B 1974 Crystal growth of deuterated KDP and determination of the deuterium content *Krist. und Techn.* **9** 87–92

Hendricks S B 1927 The crystal structure of potassium dihydrogen phosphate *Am. J. Sci.* **14** 269–87

Herring C 1951 Surface tension as a motivation for sintering *The Physics of Powder Metallurgy* ed W E Kingston (New York: McGraw-Hill) pp 143–79

Hilscher H 1985 Microscopic investigation of morphological structures on the pyramidal faces of KDP and DKDP single crystals *Cryst. Res. Technol.* **20** 1351–61

Holmberg K E 1968 *Phase Diagrams of Some Sodium and Potassium Salts in Light and Heavy Water* (Stockholm: Atomenergi)

Ivakin A A and Voronova E M 1973 Equilibriums in orthophosphoric acid solutions *Russ. J. Inorg. Chem.* **18** 885–9

Jamroz W and Górsi P 1979 Electrooptical properties of monoclinic DKDP single crystals *Phys. Status Solidi a* **53** K67–9

Jiang M, Fang Ch, Yu X, Wang M, Zheng T and Gao Z 1981 Polymorphism and metastable growth of DKDP *J. Cryst. Growth* **53** 283–91

Jona F and Shirane G 1962 *Ferroelectric Crystals* (Oxford: Pergamon)

Joshi M S and Antoni A V 1979 Nucleation in supersaturated potassium dihydrogen orthophosphate solution *J. Cryst. Growth* **46** 7–9

Kasatkin A P 1964 Effect of supersaturation on the activity of growth centers *Dokl. Akad. Sci. USSR* **154** 827–8

Kazantsev A A 1938 Solubility of potassium dihydrophosphate in water *J. Gen. Chem. (USSR)* **8** 1230–1

Kennedy N S J, Nelmes R J, Thornley F R and Rouse K D 1976 Recent structural studies of the $KH_2PO_4-KD_2PO_4$ system *Ferroelectrics* **14** 591–3

Kiehl S J and Wallance G H 1927 The dissociation pressures of monopotassium and monosodium orthophosphates and of dipotassium and disodium dihydrogen pyrophosphates. Phosphate IV *J. Am. Chem. Soc.* **49** 375–86

Kirshenbaum I 1951 *Physical Properties and Analysis of Heavy Water* (New York: McGraw-Hill)

Komukae M and Makita Y 1985 Critical slowing-down and static dielectric constant of monoclinic RbD_2PO_4 *J. Phys. Soc. Japan* **54** 4359–69

Koziejowska A and Sangwal K 1985 On the thermodynamic theory of dislocation etch-pit formation *Cryst. Res. Technol.* **20** 455–69

—— 1988 Surface micromorphology and dissolution kinetics of potassium dihydrogen phosphate (KDP) crystals in undersaturated aqueous solution *J. Mater. Sci.* **23** 2989–94

Krasil'shchikov A I 1933 Solubility of potassium acid phosphate in the presence of phosphoric acid, potassium hydroxide and potassium chloride *Ann. Inst. Anal. Phys. Chim. (Leningrad)* **6** 159–68

Krasinski M, Bartczak K and Wojchechowski B 1984 Investigation of crystal dissolution by moire fringes *Industrial Crystallization 84* ed S J Jancic and E J de Jong (Amsterdam: Elsevier) pp 241–4

Krestov G A and Abrosimov V K 1967 Effect of temperature on the 'negative' hydration of ions *J. Struct. Chem. (USSR)* **8** 822–6

—— 1972 Thermodynamics of structural changes in water during the dissolution of salts at different temperatures *Fiz. Khim. Rastvorov* ed O Ya Samoilov (Moscow: Nauka) pp 25–31

Kröger F A 1964 *The Chemistry of Imperfect Crystals* (Amsterdam: North-Holland)

Kubota N and Kubota K 1986 Secondary nucleation of magnesium sulphate from single seed crystal by fluid shear in agitated supersaturated aqueous solution *J. Cryst. Growth* **76** 69–74

Kurnakov N S 1925 Singular points of chemical diagrams *Z. Anorg. Allg. Chem.* **146** 69–102

—— 1940 *Introduction to Physicochemical Analysis* 4th edn (Moscow: Academy of Science USSR)

Kurnakov N S and Glazunov A I 1912 Alloys of cadmium with silver and copper *J. Russ. Phys. Chem. Soc.* **44** 1007–8

Kurnakov N S, Ravich M I and Troitskaya N V 1938 The ice fields of the ternary systems: base–acid–water. Formation of phosphates, chromates and borates *Ann. Secteur Anal. Phys. Chim., Inst. Chim. Gen. (USSR)* **10** 275–304

Kurnakov N S and Zhemchuzhni S F 1912 Inner friction of binary systems—characteristics of definite compounds *J. Russ. Phys. Chem. Soc.* **44** 1964–91

Kuznetsov Yu G, Chernov A A, Vekilov P G and Smol'skii I L 1987 Kinetics of growth of (101) faces of $NH_4H_2PO_4$ crystals from aqueous solution *Sov. Phys.-Crystallogr.* **32** 584–7

Landolt-Börnstein 1984 *Numerical Data and Functional Relationships in Science and Technology. New Series* vol 18 (suppl to vol III/11) ed K-H Hellwege and A M Hellwege (Berlin: Springer)

Lefaucheux F and Robert M C 1979 L'hydrodynamique, une composante importante dans la croissance des cristaux en solution *Rev. Phys. Appl.* **14** 949–59

Lemeshko G G and Egorov A N 1971 Analysis of water–heavy water on the SF-4 spectrophotometer *J. Appl. Spectrosc. (USSR)* **15** 737–8

Levich V G 1962 *Physico-chemical Hydrodynamics* (Englewood Cliffs, NJ: Prentice-Hall)

Levstik A, Blinc R, Kadaba P, Cižikov S, Levstic I and Filipič C 1975 Dielectric properties of CsH_2PO_4 and CsD_2PO_4 *Solid State Commun.* **16** 1339–41

Lilich L S and Chernykh L V 1970 Effect of the chemical nature of components on the appearance of solubility isotherm branches in ternary aqueous electrolyte solutions *Izv. Vyssh. Ucheb. Zaved., Khim. Khim. Tekhnol. (USSR)* **13** 43–8

Lilich L S, Vanyushina L N and Chernykh L V 1971 Sodium dihydrogen phosphate–phosphoric acid–water systems at 0, 25 and 50 °C *Russ. J. Inorg. Chem.* **16** 2782–6

Loiacono G M 1975 The industrial growth and characterizations of KD_2PO_4 and CsD_2AsO_4 *Acta Electron.* **18** 241–51

—— 1987 Crystal growth of KH_2PO_4 *Ferroelectrics* **71** 49–60

Loiacono G M, Balascio J F and Osborne W N 1974 Effect of deuteration on the ferroelectric transition temperature and the distribution coefficient of deuterium in $K(H_{1-x}D_x)_2PO_4$ *Appl. Phys. Lett.* **24** 455–6

Loiacono G M, Ladell J, Osborne W N and Nicolosi J 1976 Phase transitions in $Cs(D_xH_{1-x})_2AsO_4$ *Ferroelectrics* **14** 761–5

Loiacono G M, Osborne W N and Delfino M 1977 Crystal growth and properties of $Cs(D_xH_{1-x})_2AsO_4$ *J. Cryst. Growth* **41** 45–55

Loiacono G M, Zola J J and Kostecky G 1982 The taper effect in KH_2PO_4 type crystals *J. Cryst. Growth* **58** 495–9

—— 1983 Growth of KH_2PO_4 crystals of constant temperature and supersaturation *J. Cryst. Growth* **62** 545–56

Luff B B and Reed R B 1983 Enthalpies of formation, densities and heat capacities at 25 °C in the liquid-phase region of the system $K_2O–P_2O_5–H_2O$ *J. Chem. Eng. Data* **28** 39–45

Magyar H 1948 Die kristallstruktur des tetragonalen RbH_2PO_4 mit vorläufigen bemerkangen uber jene des enterspechenden Cs-salzes *Anz. Akad. Wien* **85** 166–7

Marshall W L 1982 Two-liquid-phase boundaries and critical phenomena at 275–400 °C for high-temperature aqueous potassium phosphate and sodium phosphate solution. Potential applications for steam generators *J. Chem. Eng. Data* **27** 175–80

Matthias B T 1952 Isotope effect in RbD_2PO_4 *Phys. Rev.* **85** 723–4

Matthias B T, Merz W and Scherrer P 1947 Das seignetteelektrische gitter vom KH_2PO_4 typus und das verhalten der NH_4-rotationsumwandlung bei $(NH_4, Tl)H_2PO_4$-mischkristallen *Helv. Phys. Acta* **20** 273–306

Mavrin B N, Sterin Kh E, Mishchenko A V and Rashkovich L N 1973 Low-frequency Raman spectrum and tunneling of protons (deuterons) in $Rb(D_xH_{1-x})_2PO_4$ *Sov. Phys.–Solid State* **15** 3702–5

Mersmann A and Förster W 1984 Comparison of metastable zones *Industrial Crystallization 84* ed S J Jancic and E J de Jong (Amsterdam: Elsevier) pp 245–8

Mikhailov A S, Rashkovich L N, Rzhevskii V V and Chernov A A 1989 Isotropic dislocational helix in the case of nonlinear dependence of the rate of growth steps on supersaturation *Sov. Phys.–Crystallogr.* **34** 439–45

Mishchenko A V and Rashkovich L N 1971 Crystallization of rubidium dideuterium phosphate from solutions in heavy water *Sov. Phys.–Crystallogr.* **16** 940–2

Mishchenko A V and Rashkovich L N 1973 Properties of rubidium dihydrogen phosphate solutions in normal and heavy water *Russ. J. Inorg. Chem.* **18** 2892–7

Momtaz R S and Rashkovich L N 1976 The distribution of deuterium between solid and liquid phases during the crystallization of crystals of the KDP group from aqueous solution *Phys. Status Solidi* a **38** 401–8

Monaenkova A S, Kon'kova T S, Mishchenko A V and Vorob'ev A F 1972 Enthalpies of solution of rubidium dihydrogen phosphate in water and rubidium dideuterium phosphate in heavy water *J. Gen. Chem. (USSR)* **42** 2615–9

Monaenkova A S, Vorob'ev A F, Pashlova E B, Rashkovich L N and Buznik T L 1978 Enthalpies of reciprocal transitions of monoclinic and tetragonal modifications of rubidium dihydrogen phosphate and rubidium dideuterium phosphate *J. Gen. Chem. (USSR)* **48** 3–6

Müller-Krumbhaar H, Burkhardt T W and Kroll D M 1977 A generalized kinetic equation for crystal growth *J. Cryst. Growth* **38** 13–22

Mullin J W 1972 *Crystallization* 2nd edn (London: Butterworths)

Mullin J W and Amatavivadhana A 1967 Growth kinetics of ammonium- and potassium-dihydrogen phosphate crystals *J. Appl. Chem.* **17** 151–6

Mullin J W, Amatavivadhana A and Chakraborty M 1970 Crystal habit modification studies with ammonium and potassium dihydrogen phosphate *J. Appl. Chem.* **20** 153–8

Mullin J W and Cook T P 1963 Diffusivity of ammonium dihydrogen phosphate in aqueous solution *J. Appl. Chem.* **13** 423–9

Muromtsev B A and Nazarova L A 1938 Solubility in the system ammonia–phosphoric acid–water *Bull. Acad. Sci. USSR, Ser. Chim.* N1 177–84

Nagalingam S, Vasudevan S and Ramasamy P 1981 Effect of impurities on the nucleation of ADP from aqueous solution *Cryst. Res. Technol.* **16** 647–50

Nagalingam S, Vasudevan S, Ramasamy P and Laddha G S 1980 Nucleation in quiet supersaturated ADP solution *Krist. und Techn.* **15** 1151–7

Nakano J, Shiozaki Y and Nakamura E 1973 X-ray structural study of KD_2PO_4 at room temperature *J. Phys. Soc. Japan* **34** 1423

Napijala M Lj, Žižic B, Žegarac S and Dojcilovic J 1978 The effect of pH on the growth of ammonium dihydrogen phosphate crystals *Fizika* **10** Supplement 2 502–9

Nelmes R J 1972 The crystal structure of monoclinic KD_2PO_4 *Phys. Status Solidi* b **52** K89–93

—— 1987 Structural studies of KDP and the KDP-type transition by neutron and X-ray diffraction: 1970–1985 *Ferroelectrics* **71** 83–123

Nelmes R J and Choudhary R N P 1978 Structural studies of the monoclinic dihydrogen phosphates: a neutron-diffraction study of paraelectric CsH_2PO_4 *Solid State Commun.* **26** 823–6

Nelmes R J, Eiriksson V R and Rouse K D 1972 Structural studies of the system KH_2PO_4–KD_2PO_4 *Solid State Commun.* **11** 1261–4

Newkirk H, Swain J, Stokowski S, Milam D, Smith D and Klapper D 1983 X-ray topography of laser-induced damage in potassium dihydrogen phosphate crystals *J. Cryst. Growth* **65** 651–9

Nirsha B M, Gudinitsa E N, Efremov V A, Zhadanov B V, Olikova V A and Fakeev A A 1981 Thermal dehydration of rubidium dihydrogen phosphate *Russ. J. Inorg. Chem.* **26** 2915–19

Nirsha B M, Gudinitsa E N, Fakeev A A, Efremov V A, Zhadanov B V and Olikova V A 1982 Study of the thermal dehydration of cesium dihydrogen phosphate *Russ. J. Inorg. Chem.* **27** 1366–9

Nishida Y, Yokotani A, Sasaki T, Yoshida K, Yamanaka T and Yamanaka C 1988 Improvement of the bulk laser damage threshold of potassium dihydrogen phosphate crystals by reducing the organic impurities in growth solution *Appl. Phys. Lett.* **52** 420–1

Noor J W and Dam B 1986 The growth spiral morphology on (100) KDP related to impurity effects and step kinetics *J. Cryst. Growth* **76** 243–50

Owczarek I and Wojchechowski B 1987 Nucleation and growth behavior of KDP from degassed and undegassed aqueous solutions *J. Cryst. Growth* **84** 329–31

Pakhomov V I and Sil'nitskaya G B 1970 Structure of crystals of potassium dihydrogen phosphate group ferroelectrics *Bull. Acad. Sci. USSR, Ser. Phys.* **34** 2508–10

Pastor A C 1985 Growth of ADP crystals from the melt *Mater. Res. Bull.* **20** 1165–72

Pastor A C and Pastor R C 1987 Studies in crystal growth from the melt of KDP and ADP *Ferroelectrics* **71** 61–75

Paul B K and Joshi M S 1976 Effect of supersaturation and temperature on the nucleation rate of potassium dihydrogen phosphate (KDP) crystals *Ind. J. Pure Appl. Phys.* **14** 544–6

Plyushchev V E and Stepin B D 1970 *Chemistry and Technology of Lithium, Rubidium and Cesium Compounds* (Moscow: Khimiya)

Polosin V A 1946 The polytherm of solubility in the system ammonium monophosphate, ammonium chloride, water between −15.6 °C and +35 °C *J. Phys. Chem. (USSR)* **20** 1471–4

Polosin V A and Treshchov A G 1953 The solubility polytherm in the system of urea-monoammonium phosphate in water from −15.3 to 40.0 °C *J. Phys. Chem. (USSR)* **27** 57–62

Prieto M and Amorós J L 1981 On the influence of hydrodynamic environment on crystal growth *Bull. Min.* **104** 114–19

Punin Yu O, Mirenkova T F, Artamonova O L and Ul'yanova T P 1975 Solubility of potassium dihydrogen phosphate in aqueous solutions of potassium hydroxide and phosphoric acid *Russ. J. Inorg. Chem.* **20** 2813–15

Rabinovich I B 1968 *Effect of Isotopes on Physicochemical Properties of Liquids* (Moscow: Nauka)

Rakhimov K, Zhigarnovskii B M, Polyakov Yu A, Kozhevnikov V Yu, Karapetyan F S, Takaishvili O G and Moisashvili N G 1985 The phase diagram of the CsH_2AsO_4–$CsH_{0.1}D_{1.9}AsO_4$ system *Bull. Acad. Sci. USSR, Ser. Inorg. Mater.* **21** 2095–6

Rashkovich L N 1979 Study of the cesium oxide–arsenic pentoxide–water system in the region of cesium dihydrogen arsenate crystallization at 55 °C *Russ. J. Inorg. Chem.* **24** 2806–8

—— 1984 Rapid growth of large crystals for nonlinear optics from solution *Bull. Acad. Sci. USSR* **N9** 15–19

Rashkovich L N, Koptsik V A, Volkova E N, Izrailenko A N and Plaks E M 1967 Some properties of aqueous solutions of the $NH_4H_2PO_4$ and $NH_4D_2PO_4$ *Russ. J. Inorg. Chem.* **12** 62–7

Rashkovich L N, Leshchenko V T, Amandosov A T and Koptsik V A 1983

Interferrometric investigation of growth rate of {001} faces of a triglycine sulphate crystal at various supersaturations and temperatures *Sov. Phys.–Crystallogr.* **28** 454–8

Rashkovich L N, Leshchenko V T and Sadykov N M 1982 Growth kinetics of TGS *Sov. Phys.–Crystallogr.* **27** 580–5

Rashkovich L N and Meteva K B 1978 Properties of cesium dihydrogen phosphate *Sov. Phys.–Crystallogr.* **23** 447–9

Rashkovich L N, Meteva K B and Shevchik Ya E 1977a Cesium oxide–phosphorus (V) oxide–water and cesium oxide–phosphorus (V) oxide–water D₂ systems in the region of monosubstituted cesium orthophosphate crystallization at 25 and 50 °C *Russ. J. Inorg. Chem.* **22** 1982–6

Rashkovich L N, Meteva K B, Shevchik Ya E, Hoffman V G and Mishchenko A V 1977b Growing single crystals of cesium dihydrogen phosphate and some of their properties *Sov. Phys.–Crystallogr.* **22** 613–15

Rashkovich L N, Mkrtchan A A and Chernov A A 1985 Optical interference investigation of growth morphology and kinetics of (100) face of ADP from aqueous solution *Sov. Phys.–Crystallogr.* **30** 219–23

Rashkovich L N and Momtaz R Sh 1978 Solubility of monosubstituted rubidium orthophosphate in rubidium oxide–phosphorus pentoxide–water and rubidium oxide–phosphorus pentoxide–water D₂ systems at 25 and 50 °C *Russ. J. Inorg. Chem.* **23** 1349–56

Rashkovich L N and Shekunov B Yu 1990a Morphology of growing vicinal surface. Prismatic face of ADP and KDP crystals in solutions *J. Cryst. Growth* **100** 133–44

—— 1990b Effect of hydrodynamics on ADP and KDP crystal growth in solution. I. Kinetics of growth *Sov. Phys.–Crystallogr.* **35** 160–4

—— 1990c The influence of impurities on the growth kinetics and morphology of prismatic faces of ADP and KDP crystals *Growth of Crystals* vol 18 (New York: Plenum) pp 124–39

Rashkovich L N, Shekunov B Yu and Kuznetsov Yu G 1990 Effect of hydrodynamics on ADP and KDP crystal growth in solution. II. Morphological stability of faces *Sov. Phys.–Crystallogr.* **35** 165–9

Rashkovich L N and Shustin O A 1987 Optical interferometric methods for the investigation of the crystallization kinetics in solutions *Sov. Phys.–Usp.* **30** 280–3

Ravich M I 1938 The solubility isotherm in the ternary system K₂O–P₂O₅–H₂O at 0 °C *Bull. Acad. Sci. USSR Ser. Chim.* **N1** 167–76

—— 1940 Liquidus surface of chemical diagrams *Introduction to Physicochemical Analysis* by N S Kurnakov (Moscow: Academy of Science USSR) pp 430–58

Remoissenet M, Desvignes M and Marecek V 1970 Contribution to preparation on non-wedge shaped RDP and DRDP crystals *Krist. und Techn.* **5** 535–40

Rez I S, Pakhomov V I, Sil'nitskaya G B and Fedorov P M 1967 Crystal chemistry of KH₂PO₄-type crystals *Bull. Acad. Sci. USSR Ser. Phys.* **31** 1082–5

Rilo R P and Kulikov B A 1981 Dehydration of ammonium dihydrogen phosphate (NH₄H₂PO₄) *Khim. Prom-st. (USSR)* **N12** 742–4

Robertson D S 1981 A study of the growth and growth mechanism of potassium dihydrogen orthophosphate crystals from aqueous solution *J. Mater. Sci.* **16** 413–21

Robinson A E 1949 The growth of large crystals of ammonium dihydrogen phosphate and lithium sulphate *Discuss. Faraday Soc.* **5** 315–19

Robinson R A and Stokes R H 1959 *Electrolyte Solutions* (London: Butterworths)

Rodriguez R, Aguilo M and Tejada J 1979 Unstable growth of ADP crystals *J. Cryst. Growth* **47** 518–26

Rodriguez R and Veintemillas S 1984 KDP(KH_2PO_4) growth from boiling solution *Ferroelectrics* **56** 41–4

Rosmalen R J 1977 Crystal growth processes *Thesis* University of Delft

Rosmalen R J and Bennema P 1977 The role of hydrodynamics and supersaturation in the formation of liquid inclusions in KDP *J. Cryst. Growth* **42** 224–7

Rusakov A A and Kheiker D M 1978 Refinement of the structure of RbH_2PO_4 at room temperature *Sov. Phys.–Crystallogr.* **23** 228–30

Rusakov A A, Muradyan L A and Kheiker D M 1979 Structure of deuterated rubidium dihydrogen phosphate at room temperature *Sov. Phys.–Crystallogr.* **24** 247–51

Samoilov O Ya 1957 *Structure of Aqueous Solutions of Electrolytes and Hydration of Ions* (Moscow: Academy of Science USSR)

Seidl F 1950 Concerning the seignettoelectrical behavior of RbH_2PO_4 and CsH_2PO_4 *Tscherm. Min. Petr.* **1** 432–5

Selvaratnam M and Spiro M 1965 Transference numbers of orthophosphoric acid and the limiting equivalent conductance of the H_2PO_4–ion in water at 25 °C *Trans. Faraday Soc.* **61** 360–73

Semmingsen D, Ellenson W D, Frazer B C and Shirane G 1977 Neutron-scattering study of the ferroelectric phase transition in CsD_2PO_4 *Phys. Rev. Lett.* **38** 1299–1302

Shamburov V A and Kucherova I V 1965 Variation in anomalous birefringence in crystals of KH_2PO_4 *Sov. Phys.–Crystallogr.* **10** 558–60

Shanmungham M, Gnanam F D and Ramasamy P 1984 Nucleation studies in supersaturated potassium dihydrogen orthophosphate solution and the effect of soluble impurities *J. Mater. Sci.* **19** 2837–44

—— 1985 Non-steady state nucleation process in KDP solution in the presence of XO_4 impurities *J. Mater. Sci. Lett.* **4** 746–50

Shchegrov L N, Vdovenko O P and Zhupik V I 1982 Quantitative evaluation of the composition of products of the thermal dehydration of potassium dihydrogen phosphate *Izv. Vyssh. Uchebn. Zaved. Khim. Khim. Technol. (USSR)* **25** 131–3

Shevchik Ya E, Momtaz R Sh and Rashkovich L N 1977 Isotope exchange during the crystallization of certain hydrogen containing salts from aqueous solutions *Sov. Phys.–Crystallogr.* **22** 240–2

Shimomura O and Tachibana Y 1986 Dependence of electrical conductivity of KDP solution on temperature and concentration *Trans. Inst. Electron. Commun. Eng. Japan* E **69** 471–3

Shklovskaya R M and Arkhipov S M 1967 Rubidium and cesium dihydrogen arsenates *Russ. J. Inorg. Chem.* **12** 2340–7

Singleton M F, Cooper J F, Andersen B D and Milanovich F P 1988 Laser-induced bulk damage in potassium dihydrogen phosphate crystal *Appl. Phys. Lett.* **52** 857–9

Slater J C 1941 Theory of the transition in KH_2PO_4 *J. Chem. Phys.* **9** 16–33

Smol'skii I L, Chernov A A, Kuznetsov Yu G, Parvov V F and Rozhanskii V N 1984 Vicinal sectoriality and its relation to growth kinetics of ammonium dihydrogen phosphate crystal *Dokl. Akad. Nauk USSR* **278** 358–61

—— 1985 Vicinal sectoriality in growth sectors of {011} faces of ADP crystals *Sov. Phys.–Crystallogr.* **30** 563–7

Smol'skii I L and Zaitseva N P 1991 The typical defects in fast growing KDP crystals *Growth of Crystals* vol 19 (New York: Plenum) to be published

Söhnel O 1982 Electrolyte crystal–aqueous solution interfacial tensions from crystallization data *J. Cryst. Growth* **57** 101–8

Sokolowski T 1981 Viscosity of the aqueous solutions of potassium dihydrogen phosphate (KDP) *Zesz. Nauk. Politech. Lódz. Fizyka* **6** 13–20

Sokolowski T and Kibalczyc W 1981 Conductometric studies on aqueous solution of potassium dihydrogen phosphate (KDP) *Zesz. Nauk. Politech. Lódz. Fizyka* **6** 25–37

Stephenson C C, Corbella J M and Russell L A 1953 Transition temperatures in some dihydrogen and dideutero phosphates and arsenates and their solid solutions *J. Chem. Phys.* **21** 1110

Strel'nikova R V and Rashkovich L N 1977 Ranges of existence of the monoclinic and tetragonal forms $K(D_xH_{1-x})_2PO_4$ in aqueous solutions *Sov. Phys.–Crystallogr.* **22** 483–5

Strickland-Constable R F 1968 *Kinetics and Mechanism of Crystallization* (London: Academic)

Sumita M, Osaka T and Makita Y 1981 New phase transitions of monoclinic RbD_2PO_4 and its forced transition to the ferroelectric state *J. Phys. Soc. Japan* **50** 154–8

Swain J E, Stokowski S E, Milam D and Kennedy G C 1982 The effect of baking and pulsed laser irradiation on the bulk laser damage threshold of potassium dihydrogen phosphate crystals *Appl. Phys. Lett.* **41** 12–14

Synowiec J 1973 A method of calculation of the limiting supersaturation of inorganic salt solutions *Krist. und Techn.* **8** 701–8

Takubo H 1985 Dependence of growth rate of solution velocity in growth of $NH_4H_2PO_4$ crystals *J. Cryst. Growth* **72** 631–8

Takubo H, Kume Sh and Koizumi M 1984 Relationships between supersaturation, solution velocity, crystal habit and growth rate in crystallization of $NH_4H_2PO_4$ *J. Cryst. Growth* **67** 217–26

Takubo H and Makita H 1989 Refractometric studies of $NH_4H_2PO_4$ and KH_2PO_4 solution growth; experimental setup and refractive index data *J. Cryst. Growth* **94** 469–74

Thornley F R and Nelmes R J 1975 A neutron diffraction study of room temperature monoclinic KD_2PO_4 *J. Chem. Phys. Lett.* **34** 175–7

Torgesen J L and Jackson R W 1965 Growth layers on ammonium dihydrogen phosphate *Science* **148** 952–4

Tsukamoto K 1983 In situ observation of mono-molecular growth steps of crystals growing in aqueous solution *J. Cryst. Growth* **61** 199–209

Ubbelohde A R 1939 Structure and thermal properties associated with some hydrogen bonds in crystals. II. Further examples of isotope effect *Proc. R. Soc.* A **173** 417–27

Ubbelohde A R and Woodward I 1939 Isotope effect in potassium dihydrogen phosphate *Nature* **144** 632

—— 1942 Structure and thermal properties associated with some hydrogen bonds in crystals. III. Isotope effect in some acid phosphates *Proc. R. Soc.* A **179** 399–407

Uesu Y and Kobayashi J 1976 Crystal structure and ferroelectricity of cesium dihydrogen phosphate CsH_2PO_4 *Phys. Status Solidi* a **34** 475–81 van der Erden J P and Müller-Krumbhaar H 1986 Dynamic co-arsening of crystal surfaces by formation of macrosteps *Phys. Rev. Lett.* **57** 2431–3

van Enckevort W J P 1984 Surface interferometry of aqueous solution grown crystals *Prog. Cryst. Growth Charact.* **9** 1–50

van Enckevort W J P, van Rosmalen R J, Klapper H and van der Linden W H 1982 Growth phenomena of KDP crystals in relation to the internal structure *J. Cryst. Growth* **60** 67–78

van Enckevort W J P, van Rosmalen R J and van der Linden W H 1980 Evidence for spiral growth on the pyramidal faces of KDP and ADP single crystals *J. Cryst. Growth* **49** 502–14

Vannay L 1987 Growth of potassium dihydrogen phosphate (KDP) single crystals by means of thermoelectric cooler *Acta Phys. Hung.* **61** 197–200

van Wazer J R 1958 *Phosphorus and its Compounds* vol 1 (New York: Interscience)

Varshavskii Ya M and Vaisberg S E 1957 Thermodynamic and kinetic peculiarities of isotope exchange reactions of hydrogen *Usp. Khim. (USSR)* **27** 1434–68

Vasilevskaya A S, Volkova E N, Koptsik V A, Rashkovich L N, Regul'skaya T A, Rez I S, Sonin A S and Suvorov V S 1967 Nonlinear optical properties of single crystals of deuterated ammonium dihydrogen phosphate *Sov. Phys.–Crystallogr.* **12** 446–7

Veintemillas S, Lefaucheux F and Robert M C 1987 X-ray topographic study of KH_2PO_4 crystals grown from boiling solutions *J. Cryst. Growth* **80** 289–97

Vogel L, Figurski G and Vohland P 1983 Untersuchungen über das löslichkeits-verhalten von electrolyten und nichteelectrolyten in binaren systemen *Z. Chem.* **23** 331–2

Volfkovich S I, Berlin L E and Mantzev B M 1932 Physicochemical investigation of the process for manufacturing ammonium phosphates (ammophos) *J. Appl. Chem. (USSR)* **5** 1–14

Volkova E N, Berezhnoi B M, Izrailenko A N, Mishchenko A V and Rashkovich L N 1971a Electrooptical and optical properties of partially deuterated rubidium dihydrogen phosphate crystals *Bull. Acad. Sci. USSR, Ser. Phys.* **35** 1858–61

Volkova E N, Dianova V A, Zuev A L, Izrailenko A N, Lipatov A S, Parygin V N, Rashkovich L N and Chirkov L E 1971b Electrooptic and piezoelectric properties of α-HIO_3 crystals *Sov. Phys.–Crystallogr.* **16** 284–6

Volkova E N, Podshivalov Yu S, Rashkovich L N and Strukov B A 1975 Effect of deuterium concentration on the Curie point of some potassium dihydrogen phosphate group crystals *Bull. Acad. Sci. USSR, Ser. Phys.* **39** 787–90

Vorob'ev A F, Monaenkova A S, Mishchenko A V and Rashkovich L N 1974 Thermodynamics of rubidium dihydrogen phosphate and rubidium dideuterium phosphate in water and heavy water *Izv. Vyssh. Ucheb. Zaved. Khim. Khim. Tekhn. (USSR)* **17** 673–6

Watkins Ch and Jones H C 1915 Conductivity and dissociation of some rather unusual salt in aqueous solution *J. Am. Chem. Soc.* **37** 2626–36

Wenyu J 1985 The ammonia–phosphoric acid–water system at 100 °C *J. Chengdu Univ. Sci. Technol.* N2 41–3

West J 1930 A quantitative X-ray analysis of the structure of potassium dihydrogen phosphate (KH_2PO_4) *Z. Kristallogr.* **74** 306–32

Wilke K-Th 1963 Methoden der Kristallzüchtung (Berlin: VEB)

Wissing R, Elwenspoek M and Degens B 1986 In situ observation of secondary nucleation *J. Cryst. Growth* **79** 614–19

Wojciechowski K and Kibalczyc W 1986 Light scattering study of KH_2PO_4 and $BaSO_4$ nucleation process *J. Cryst. Growth* **76** 379–82

Yakushkin E D and Anisimova V N 1988 Phase transition in monoclinic KD_2PO_4 crystal *Phys. Status Solidi* a **105** 139–43

Yokotani A, Fujoka K, Nishida Y, Sasaki T, Yamanaka T and Yamanaka Ch 1987 Formation of macroscopic hillocks of the prismatic faces of KDP crystals due to microbes in the solution *J. Cryst. Growth* **85** 549–52

Yokotani A, Koide H, Sasaki T and Yamanaka T 1984 Fast growth of KDP single crystals by electrodialysis method *J. Cryst. Growth* **67** 627–32

Yokotani A, Koide H, Yamamuro K, Sasaki T, Yamanaka T and Yamanaka Ch 1983 Solution growth of large KDP crystals by rotating fluid method *Tech. Rep. Osaka Univ.* **33** 301–8

Yokotani A, Sasaki T, Yoshida K, Yamanaka T and Yamanaka Ch 1986 Improvement of the bulk laser damage threshold of potassium dihydrogen phosphate crystals by ultraviolet irradiation *Appl. Phys. Lett.* **48** 1030–2

Zaitseva N P 1989 Fast growth of KDP and DKDP crystals from high supersaturated aqueous solutions *Thesis* Moscow University

Zhigarnovskii B M, Polyakov Yu A, Bugakov V I, Maifat M A, Rakhimov K, Moisashvili N G, Takaishvili O G, Mdinaradze A G and Orlovskii V P 1984 Physicochemical studies of potassium, rubidium and cesium hydrogen phosphates and arsenates *Bull. Acad. Sci. USSR, Ser. Inorg. Mater.* **20** 1243–8

Zvorykin A Ya and Ratnikova V D 1963 Isothermal solubility in the system CsH_2PO_4–$NH_4H_2PO_4$–H_2O at 25 °C *Russ. J. Inorg. Chem.* **8** 1018–19

Zvorykin A Ya and Vetkina L S 1961 Isothermal solubility in the system RbH_2PO_4–$NH_4H_2PO_4$–H_2O at 25 °C *Russ. J. Inorg. Chem.* **6** 2572–5

Zwicker B and Scherrer P 1944 Elektrooptische eigenschaften der seignette-elektrischen kristalle KH_2PO_4 und KD_2PO_4 *Helv. Phys. Acta* **17** 346–73

Index

Milton Keynes UK
Ingram Content Group UK Ltd.
UKHW040058071024
449327UK00019B/646